Serdar Öztekin
Milan Martinov
Editors

Medicinal and Aromatic Crops
Harvesting, Drying, and Processing

Pre-publication REVIEWS, COMMENTARIES, EVALUATIONS . . .

"I found the drying and industrial usage chapters extremely valuable and informative in view of their direct application in practice. I was impressed by the convincing review of the economic aspects of agricultural production of medicinal and aromatic crops, as well as their application in cosmetics, perfumery, medicine and other branches, with the help of modern user-friendly software."

Nicolay Mihailov, PhD
Dean of the Faculty of Electrical and Electronic Engineering, "Angel Kunchev" University of Rousse, Bulgaria

"The structure of the book is reader friendly and it provides information on a wide range of topics within medicinal and aromatic crops. Especially useful are the comprehensive chapters on harvesting, drying, and dry processing, which give a closer look into a diversity of techniques and procedures and even tips and tricks for growers! This is not only a compilation of knowledge gathered during years of university science work; it is an extensive, useful, and applicable guide through the vibrant world of medicinal and aromatic crops."

Peter Schulze Lammers, PhD
Professor of Technology in Crop Farming at the University of Bonn, Germany

More pre-publication
REVIEWS, COMMENTARIES, EVALUATIONS . . .

"This book provides good information and ideas for developers, supported by a good number of schemes and photographs. *Medicinal and Aromatic Crops* covers a domain not common in literature and is useful for practitioners, engineers, and producers. In fact, not only does this book review the know-how relative to a variety of medicinal and aromatic crops and related uses, but it also provides useful ideas for adapting existing machines or developing new ones for various operations from harvesting to final processing, as well as to support the respective decision making. The sections on harvesting, drying, and dry processing are particularly useful for agricultural and mechanical engineers and other related professionals."

Luis Santos Pereira, PhD
Professor, Institute of Agronomy, Technical University of Lisbon, Portugal; President, International Commission of Agricultural Engineering, CIGR

"This book is well balanced. Discussions on the advantages and disadvantages of the methods for each production phase are based on the tremendous experience of the authors and provide excellent material to enlighten producers and processors. *Medicinal and Aromatic Crops* is very impressive. Useful recommendations are made on natural, mechanical, and microwave drying, practical approaches to dryer selection, and energy costs, process control, product moisture content control, as well as the use of renewable energy sources including solar energy, and the use of solid biomass. All these are elaborated in an elegant fashion."

Shlomo Navarro, PhD
Professor Emeritus, Department of Food Science, Agricultural Research Organization, the Volcani Center, Bet Dagan, Israel

HFAPP

Haworth Food & Agricultural Products Press™
An Imprint of The Haworth Press, Inc.
New York

NOTES FOR PROFESSIONAL LIBRARIANS AND LIBRARY USERS

This is an original book title published by Haworth Food & Agricultural Products Press™, an Imprint of The Haworth Press, Inc. Unless otherwise noted in specific chapters with attribution, materials in this book have not been previously published elsewhere in any format or language.

CONSERVATION AND PRESERVATION NOTES

All books published by The Haworth Press, Inc., and its imprints are printed on certified pH neutral, acid-free book grade paper. This paper meets the minimum requirements of American National Standard for Information Sciences-Permanence of Paper for Printed Material, ANSI Z39.48-1984.

DIGITAL OBJECT IDENTIFIER (DOI) LINKING

The Haworth Press is participating in reference linking for elements of our original books. (For more information on reference linking initiatives, please consult the CrossRef Web site at www.crossref.org.) When citing an element of this book such as a chapter, include the element's Digital Object Identifier (DOI) as the last item of the reference. A Digital Object Identifier is a persistent, authoritative, and unique identifier that a publisher assigns to each element of a book. Because of its persistence, DOIs will enable The Haworth Press and other publishers to link to the element referenced, and the link will not break over time. This will be a great resource in scholarly research.

Medicinal and Aromatic Crops
Harvesting, Drying, and Processing

HAWORTH FOOD & AGRICULTURAL PRODUCTS PRESS™
Crop Science

The Lowland Maya Area: Three Millennia at the Human-Wildland Interface edited by A. Gómez-Pompa, M. F. Allen, S. Fedick, and J. J. Jiménez-Osornio

Biodiversity and Pest Management in Agroecosystems, Second Edition by Miguel A. Altieri and Clara I. Nicholls

Plant-Derived Antimycotics: Current Trends and Future Prospects edited by Mahendra Rai and Donatella Mares

Concise Encyclopedia of Temperate Tree Fruit edited by Tara Auxt Baugher and Suman Singha

Landscape Agroecology by Paul A. Wojtkowski

Concise Encyclopedia of Plant Pathology by P. Vidhyasekaran

Molecular Genetics and Breeding of Forest Trees edited by Sandeep Kumar and Matthias Fladung

Testing of Genetically Modified Organisms in Foods edited by Farid E. Ahmed

Fungal Disease Resistance in Plants: Biochemistry, Molecular Biology, and Genetic Engineering edited by Zamir K. Punja

Plant Functional Genomics edited by Dario Leister

Immunology in Plant Health and Its Impact on Food Safety by P. Narayanasamy

Abiotic Stresses: Plant Resistance Through Breeding and Molecular Approaches edited by M. Ashraf and P. J. C. Harris

Teaching in the Sciences: Learner-Centered Approaches edited by Catherine McLoughlin and Acram Taji

Handbook of Industrial Crops edited by V. L. Chopra and K. V. Peter

Durum Wheat Breeding: Current Approaches and Future Strategies edited by Conxita Royo, Miloudi M. Nachit, Natale Di Fonzo, José Luis Araus, Wolfgang H. Pfeiffer, and Gustavo A. Slafer

Handbook of Statistics for Teaching and Research in Plant and Crop Science by Usha Rani Palaniswamy and Kodiveri Muniyappa Palaniswamy

Handbook of Microbial Fertilizers edited by M. K. Rai

Eating and Healing: Traditional Food As Medicine edited by Andrea Pieroni and Lisa Leimar Price

Physiology of Crop Production by N. K. Fageria, V. C. Baligar, and R. B. Clark

Plant Conservation Genetics edited by Robert J. Henry

Introduction to Fruit Crops by Mark Rieger

Generations Gardening Together: Sourcebook for Intergenerational Therapeutic Horticulture by Jean M. Larson and Mary Hockenberry Meyer

Agriculture Sustainability: Principles, Processes, and Prospects by Saroja Raman

Introduction to Agroecology: Principles and Practice by Paul A. Wojtkowski

Handbook of Molecular Technologies in Crop Disease Management by P. Vidhyasekaran

Handbook of Precision Agriculture: Principles and Applications edited by Ancha Srinivasan

Dictionary of Plant Tissue Culture by Alan C. Cassells and Peter B. Gahan

Handbook of Potato Production, Improvement, and Postharvest Management edited by Jai Gopal and S. M. Paul Khurana

Carbon Sequestration in Soils of Latin America edited by Rattan Lal, Carlos C. Cerri, Martial Bernoux, Jorge Etchevers, and Eduardo Cerri

Medicinal and Aromatic Crops: Harvesting, Drying, and Processing edited by Serdar Öztekin and Milan Martinov

Mycorrhizae in Crop Production edited by Chantal Hamel and Christian Plenchette

Integrated Nutrient Management for Sustainable Crop Production edited by Milkha S. Aulakh and Cynthia A. Grant

Asian Crops and Human Dietetics by Usha Rani Palaniswamy

Medicinal and Aromatic Crops
Harvesting, Drying, and Processing

Serdar Öztekin
Milan Martinov
Editors

Haworth Food & Agricultural Products Press™
An Imprint of The Haworth Press, Inc.
New York

For more information on this book or to order, visit
http://www.haworthpress.com/store/product.asp?sku=5895

or call 1-800-HAWORTH (800-429-6784) in the United States and Canada
or (607) 722-5857 outside the United States and Canada

or contact orders@HaworthPress.com

Published by

Haworth Food & Agricultural Products Press™, an imprint of The Haworth Press, Inc., 10 Alice Street, Binghamton, NY 13904-1580.

PUBLISHER'S NOTE
The development, preparation, and publication of this work has been undertaken with great care. However, the Publisher, employees, editors, and agents of The Haworth Press are not responsible for any errors contained herein or for consequences that may ensue from use of materials or information contained in this work. The Haworth Press is committed to the dissemination of ideas and information according to the highest standards of intellectual freedom and the free exchange of ideas. Statements made and opinions expressed in this publication do not necessarily reflect the views of the Publisher, Directors, management, or staff of The Haworth Press, Inc., or an endorsement by them.

Cover design by Jennifer M. Gaska.
Cover photo by Milan Martinov.

Library of Congress Cataloging-in-Publication Data

Medicinal and aromatic crops : harvesting, drying, and processing / Serdar Öztekin, Milan Martinov, editors.
p. cm.
Includes bibliographical references.
ISBN: 978-1-56022-974-2 (case : alk. paper)
ISBN: 978-1-56022-975-9 (soft : alk. paper)
1. Aromatic plants. 2. Medicinal plants. I. Öztekin, Serdar. II. Martinov, Milan.

SB301.M43 2007
635'.7—dc22

2006022949

CONTENTS

About the Editors **ix**

Contributors **xi**

Preface **xiii**

Chapter 1. Introduction **1**
Serdar Öztekin
Milan Martinov

Characteristics of Medicinal and Aromatic Plant Production 1
Mechanization of MAP Production 19

Chapter 2. Harvesting **27**
Milan Martinov
Miodrag Konstantinovic

Introduction 27
Manual and Semimechanized Harvesting 29
Mechanized Harvesting 37
Transport 81

Chapter 3. Drying **85**
Milan Martinov
Serdar Öztekin
Joachim Müller

Introduction 85
Hot Air Drying Parameters 97
Hot Air Dryers 103
Implementation of Renewable Energy Sources 119

Chapter 4. Mechanical Processing **131**
Milan Martinov
Miodrag Konstantinovic

Introduction 131
Predrying Processes 132
Postdrying Processes 142

Chapter 5. Extraction **211**
Yurtsever Soysal
Serdar Öztekin

Introduction 211
Conventional Extraction Techniques 213
New Extraction Techniques 236

Chapter 6. Industrial Utilization **253**
Yurtsever Soysal
Serdar Öztekin

Introduction 253
Medicinal and Pharmaceutical Purposes 254
Food and Food Ingredients 256
Herbal Tea 257
Cosmetics, Perfumery, and Aromatherapy 258
Pest and Disease Control 260
Nonconventional Uses 261

Chapter 7. Decision Making **267**
Sait M. Say

Introduction 267
Decision-Making Procedures 269
Factors for New Enterprise Decision 270
Financial Analysis 282
Software for Decision Making 296

Appendix: User-Friendly Software for Decision Making: FANE v 1.0 **299**

Start Screen 299
New Project Screen (Data Entry Screen) 300
General Screen 300
Investment Inputs Screen 301
Seasonal Inputs Screen 301
Revenue/s Screen 302
Decision Parameters Screen 303
Results Screen 305

Index **309**

ABOUT THE EDITORS

Serdar Öztekin, PhD, has been Professor of Agricultural Engineering at Cukurova University in Adana, Turkey, since 1990. His research interests are in the areas of post-harvest processing and the storage of agricultural crops. Dr. Öztekin has published five books on agricultural mechanization, has designed and built five machines for different crop processing stages, and was involved in the foundation of the Association of Agricultural Engineering in Southeastern Europe.

Milan Martinov, PhD, is Professor of Agricultural Engineering at the University of Novi Sad in Serbia. His research interests include the development of procedures and machines for special crops, the implementation of renewable energies, and the development of rural areas. Dr. Martinov is Editor-in-Chief of the regional journal, *Agricultural Engineering—Reports for Southeastern Europe,* Associate Editor of "CIGR," the e-journal for the World Organization of Agricultural Engineering, and is an editorial board member for several other agricultural engineering journals.

doi:10.1300/5895_a

CONTRIBUTORS

Mr. Miodrag Konstantinovic, MSc, born in 1974, completed graduate and postgraduate studies in mechanical engineering with an emphasis on agricultural mechanization at the University of Novi Sad. He is presently in the final stage of PhD studies at the Institute of Agricultural Engineering in Bonn, Germany. Mr. Konstantinovic started his activities in this innovative field in 1999, during the last phase of graduate studies that dealt with a machine line for Dog Rose processing. In the course of postgraduate studies he engaged in several R&D projects dealing with special and intensive crops. The problems and possibilities of improvement of broomcorn seed processing was the subject of his master's thesis. He has been the author and co-author of numerous professional and scientific publications.

Dr. Joachim Müller, born in 1959, graduated in agricultural engineering at the University of Hohenheim (Stuttgart, Germany) where he also did his PhD research on drying of medicinal plants. After a professorship in farm technology at Wageningen University (The Netherlands) he returned to the University of Hohenheim, where he currently holds the chair of agricultural engineering in the tropics and subtropics. In interdisciplinary research projects, Dr. Müller is focusing on value-adding post-harvest processing and application of renewable energy, where processing of medicinal plants takes a prominent position. The results have been published in more than fifty journal articles and a couple of his technological developments are patented. Dr. Müller is a member of the editorial boards of several scientific journals and co-editor of "Zeitschrift für Arznei-und Gewürzpflanzen."

Dr. Sait M. Say is currently an assistant professor in the agricultural faculty of Cukurova University where he started his professional career as a research assistant in the Farm Machinery Department in 1994. He completed his PhD study in 2001 at the same department.

Medicinal and Aromatic Crops

doi:10.1300/5895_b

He has conducted research in the fields of agricultural mechanization management, engineering economics in agriculture, and precision farming. He is the author or co-author of twenty research papers in various congresses. In addition, two papers have been published in scientific refereed journals.

Dr. Yurtsever Soysal is currently an associate professor of agricultural engineering at Mustafa Kemal University, Antakya, Hatay Turkey, and is the head of the Agricultural Machinery Department. Earlier, he was a research and teaching assistant of the Agricultural Machinery Department at Cukurova University in Adana, Turkey. He earned his BS (1996) and PhD (2000) degrees from Cukurova University. He is the author or co-author of nearly fifty professional papers focused primarily on drying of medicinal and aromatic plants.

Preface

There is an obvious rise in the market and production of medicinal and aromatic plants around the world. This is accompanied by enormous activities of R&D institutions, workshops, conferences, publications, books, and the like. These activities are aimed to support the sector up to the market-ready products and business development. The huge number of publications in this field is related to the breeding, growing, processing technology, safety and quality, environmental and social impacts, production and marketing management, and the like of medicinal and aromatic plants. There is one field that is only very slightly covered—that is the field of mechanization. The question without an answer is—why? Appropriate technology and profitable production can be realized only by fully or mostly mechanized processes. That is the case even in developing countries faced with worldwide competition. Mechanization contributes not only to the rise of productivity but enables uniform quality and reduction of drudgery. Some publications include sections related to the mechanization of growing, harvesting, and processing, but without enough details, giving only general instructions, demands, and descriptions. That is also the case of the most modern information source, the Internet. What are the reasons for this? Some of them follow:

1. Development of mechanization is not a typical scientific discipline, and researchers are not interested in involving themselves in this rather complicated field without prospect for recognition. Labor and cost-intensive activities cannot bring very attractive

Medicinal and Aromatic Crops

doi:10.1300/5895_c

outcomes in comparison with other fields. It is also more difficult to get adequate funds.
2. Every new machinery development is expensive. The sold products should recoup the costs of development. The market and customer demand for medicinal and aromatic plants machinery, that is, special machinery, is small. Because of this, the developed machines are expensive. That is why "big" industry is not interested in producing the special machines required.
3. Wild crafting and growing of medicinal and aromatic plants is today practiced mostly in developing countries. The farmers do not have the capacity to purchase high-tech machines, the use of which would be, with few exceptions, inadequate and not profitable. The use of simple, traditional devices enables successful production with low input in most cases. Nevertheless, the improvement of mechanization is needed, but under these conditions a special approach is necessary.
4. The high diversity of medicinal and aromatic plants make it impossible to develop devices that could be universally used, or at least for a large group of species. Sometimes a redesigned machine, originally aimed for other crops, can be successfully used for medicinal and aromatic plants, but it requires some improvements to the machine. For some plants it is very difficult to provide appropriate mechanization due to the superior quality of manual work. This could be solved, but close cooperation with R&D staff in other fields, for example, breeding and agrotechnology development, is needed.

Soil tillage, seeding, and crop protection of medicinal and aromatic plants could be solved in most cases by the use of universal mechanization. This book presents the mechanization for harvesting and primary processing, up to the procedures that provide market-ready products. Some general data on medicinal and aromatic plants, contemporary trends in documented production, and instructions for setting up or improving production processes are given. The book is also aimed to initiate the reader's creativity and to inspire them to step into development and innovations. The book is based on the authors' experiences and on a survey of published information, including the Internet.

The book is intended for producers of medicinal and aromatic plants, all professionals involved in the field, experienced and beginners, and it can be used as a teaching material for students.

Hopefully, the book will compensate for the unjustifiable lack of publications on mechanization of medicinal and aromatic plant production and contribute to further development of the profession and business of this wonderful human activity.

Chapter 1

Introduction

Serdar Öztekin
Milan Martinov

CHARACTERISTICS OF MEDICINAL AND AROMATIC PLANT PRODUCTION

Medicinal plants have been used for different purposes in many regions of the world since ancient times. After World Health Organization (WHO), medicinal plants are commonly used in preventing and treating specific ailments and diseases and are generally considered to play a beneficial role in health care (Srivastava et al., 1996). Some cultivars from medicinal plant families are also used as ingredients to season or to give a pleasant flavor or smell to foods. Therefore, the terms "medicinal" and "aromatic" are usually used in conjunction. In this book, MAP is preferred as the acronym of medicinal and aromatic plants.

The general classification of MAP is given in Table 1.1. Terms such as "herbs" and "spices" are often used interchangeably, both in practice and in different sources. There is really no standard definition for the term herb, and the distinction between herbs and spices is not very clear. Herb comes from the Latin "herba" meaning green crops. It refers to virtually all cultivated and wild plants, especially to their green, succulent parts (Anonymous, 1992). To the botanist, a herb is a nonwoody plant that dies back to the ground in winter. For herbalists, a herb is any plant or plant part that has useful properties beyond being a garden ornament. In this broad sense, herbs are plants

Medicinal and Aromatic Crops

doi:10.1300/5895_01

TABLE 1.1. Classification of MAP according to different criteria.

Criteria	Sample plant
Family	Acanthaceae *(Justica achatoda),* Labitae (*Melisa officinalis*)
Active ingredients	Essential oil *(Rosa damescena),* Bitter substances *(Artemisia absinthium),* Glycosides *(Urginea maritime),* Saponins *(Saponaria officinalis),* Alkaloids *(Cinchona pubescens),* Flavonoids *(Calendula officinalis)*
Usage	Plants for drinks, herbal tea *(Matricaria chamomilla),* spices *(Thymus vulgaris),* medicinal plants *(Papaver somniferum),* cosmetic plants *(Lavendula officinalis)*
Used plant part	Rhizome *(Zingiber officinale),* roots *(Panax ginseng),* bulbs *(Allium sativum),* barks *(Hydrangea arborescens),* leaves *(Mentha piperita),* herbarium *(Artemisia Absinthium),* flowers *(Rosa damescena),* seeds and fruits *(Pimpinella anisum)*
Pharmacological influence	Neural system regulator *(Apium graveolens),* blood system regulator *(Ginseng),* carminative-stomachic *(Matricaria chamomilla),* diuretic *(Petroselinum sativum)*

Source: Compiled from Ceylan, 1987, 1995.

used for flavoring foods and beverages; for medicines, cosmetics, dyes, and perfumes; and for other household and economic uses.

Specific herbs may be valued for their leaves (such as basil and parsley), flowers (chamomile), seeds (dill), stems (angelica), or underground tissues (garlic). Generally, the concept of herbs as flavoring agents excludes those plants commonly known as vegetables and is further limited to plants grown in temperate regions. A contrasting group of economic flora consists of the spice plants. These are usually understood to be plants from the tropics that bear aromatic fruits, seeds, or woody barks and are used today primarily for seasoning foods. Examples of popular culinary spices include cinnamon, ginger, and black pepper.

MAP are the oldest known health care products in traditional medicine, which existed before the arrival of modern medicine, and are still used as such in many countries. Other than their traditional uses

MAP are valuable raw crops for manufacturing different kinds of industrial products (Figure 1.1). MAP are mainly used as

- fresh material,
- dried, conserved material,
- thermo-chemically processed material, and
- industrially finalized material.

Specific characteristics of medicinal and aromatic plants production could be given as follows:

- There is a huge diversity of plants and species.
- Active ingredients have crucial importance.
- Agro-ecological conditions have more impact on active ingredients.
- The market is relatively small and sensitive, prices are instable, and there is worldwide competition.
- Specific production procedures and equipment are required.

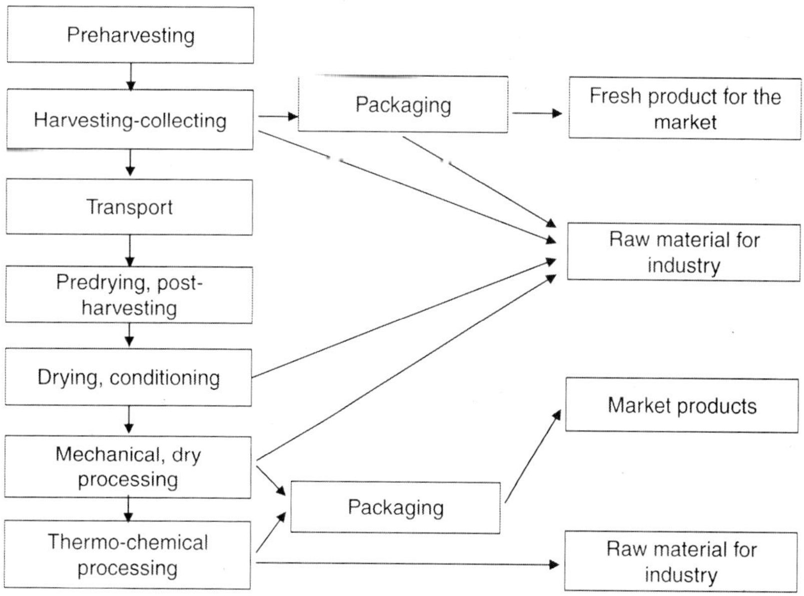

FIGURE 1.1. Flow chart for MAP products.

Although many plants are used traditionally for medicinal and aromatic purposes, most of them are still not registered. Worldwide, out of an estimated 250,000 to 500,000 higher plant species identified so far, about 35,000 (some estimate up to 75,000) plants are used for medicinal purposes (Table 1.2) (Lewington, 1993; Comer and Debry, 1996; Baser, 1997; Nair and Ganapathi, 1998). Only about 100 plants are regularly cultivated, whereas the remaining are collected in their natural habitat. For species gathered from spontaneous flora the terms "wild crafting" or "wild harvesting" are used.

MAP are mainly produced to extract the active ingredients that are created during secondary material exchange, and these are used for pharmaceutical purposes. Without active ingredients MAP are not different from forage. The concentration of active ingredients in MAP is relatively low, usually within the range of 0.1 to 2 percent of dry matter. The properties of the active ingredients are influenced by processing procedures.

To control the influence of agro-ecological conditions on active ingredients, MAP should be cultivated. Wild crafting does not help reach the required quality and quantity of active ingredients. Even today some herbs continue to be collected from spontaneous flora without any management or systematization of parameters that allow the quality control of the whole processing. Until the past few years,

TABLE 1.2. Number of MAP used in some countries.

Country	Number of species	References
China	5,000-9,905	Lambert et al., 1997; Kuipers, 1995
India	2,500-8,000	Lambert et al., 1997; Nair and Ganapathi, 1998
European Countries	> 2,000	Lange, 1998
Peninsular Malaysia and Neighboring Islands	6,000-7,000	Nair and Ganapathi, 1998
Sri Lanka	1,400	Nickel and Sennhauser, 2001
Nepal	700	Nickel and Sennhauser, 2001
Colombian Amazon	2,000	Anonymous, 1997
Turkey	> 500	Koyuncu, 1995
Tunisia	152	Heyvood, 1999
Iraq	> 1,500	Heyvood, 1999
Total World	35,000-75,000	Baser, 1997; Nair and Ganapathi, 1998

many producers preferred wild crafting because of the natural availability of the material and the relatively low labor costs. It also facilitated the collection of the required quantity. But in many countries new legislative regulations and standards demand full proof of quality (Bomme, 1992). To get the official permission for producing MAP-based products producers should provide appropriate records on the growing conditions of the plant. The homogeneity of plant material, the best proper active ingredients, minimum contamination by heavy metals, pesticides, nitrate, radioactive substances, and microorganisms, and the presence of any foreign material or the mixing with unclean or infected plant parts are defined in detail in the codes and strictly followed by governmental or nongovernmental organizations. These high demands could be fulfilled only by carefully planned and controlled cultivation. Wild collected raw material cannot meet the stipulations with regard to unknown contamination, falsification, or site-specific differences. On the other hand, the survival of many species in their natural habitat is under growing threat due to improper collecting. The protection of plant diversity is also an important aspect that makes cultivation necessary. In addition to these environmental problems spontaneous flora cannot ensure a regular supply for market demand, which requires certain volumes. Therefore, the quality and quantity of wild material cannot meet the requirements of a market that is getting stringent in enforcing related standards. This situation creates ample opportunities for countries with ecologically suitable conditions to derive economic benefit from MAP cultivation. It provides variety to maintain the genetic diversity of each species, to select new varieties and create optimum growing conditions, and to allow the use of improved farming methods.

When one decides to do MAP cultivation the foremost requirements are information on market situation, reproducing material, and production (Davis, 2001). Market oriented-cultivation without considering the specific demands of MAP may lead to unsuccessful results. The importance of the marketability of cultivated crops does not require additional emphasis. This point is especially crucial for MAP because in most producer countries there is no free market for fresh or conserved crops. Even if the domestic market is available, the international market grows much faster. However, the international market for MAP is relatively small and very sensitive, depending on production, consumer habits, and the like.

The production of many MAP species is usually concentrated in only one or, at the most, a few countries. Any problem that could drastically reduce the yield has a great influence on the market prices. This means that if any country is not able to offer the crop for the world market, then there is a serious shortage, because it is not possible to compensate from other producers; in which case the price of the crop will increase. Or else, if a new producer offers a crop to the international market, the price will decrease. This situation is not typical for other field and horticultural crops, because there are many producers from different countries. On the other hand, the consumer habits are strongly influenced by the audiovisual press. Although there is no reliable scientific evidence that supports the use of any MAP-based product, the demand for the product could increase in an unpredictable manner in some years. Perhaps this behavior could be defined as consumer fashion.

Production Steps and Procedures

Agriculture has different roles in developed and developing countries. In developed countries it is usually maintained to protect the tradition of society. The agricultural crop requirement is usually covered by import from producer countries where agriculture plays a more important role in the economy. But, mass production and improper farming methods in producer countries have been the cause of many environmental problems. The disturbing of the natural equilibrium and environmental pollution are often discussed on different platforms. In this context, MAP could open up a new perspective for individual small and medium enterprises. Even experienced farmers require additional knowledge on cultivation and marketing for the growing and processing of MAP. They should follow new developments to refresh their agricultural background.

To cover the large information demand for MAP cultivation new possibilities such as ICT (information and communication techniques) could also offer a different way for communicating and exchanging information and knowledge. At the local level, ICT provide growers with much information on cultivation practices, market prices, and social services, such as health and knowledge and education (Accascina, 2000). Some examples of areas where ICT could play a catalytic role in developing rural areas are given by Munyua (2000):

- Decision-making process
- Market outlook
- Empowering rural communities
- Targeting marginalized groups
- Creating employment

Safety and Quality

International and national standards on foods and other documents have been developed to protect the health of consumers and facilitate the trade in food. One of the most important standards is the *Codex Alimentarus,* which means nutrition code in Latin, prepared with the initiative of WHO and FAO (Anonymous, 2003-II). *Codex Alimentarus* covers all the main foods, whether processed, semi-processed, or raw. It defines the hygienic and nutritional quality of food, including microbiological norms, food additives, pesticide and veterinary drug residues, contaminants, labeling and presentation, methods of sampling, and risk analysis. WHO has also recognized the importance of Hazard Analyses and Critical Control Points (HACCP) system, which focuses on preventing food safety hazards, relies more heavily on scientific principles, permits more efficient government oversight, and places greater responsibility on food operations to ensure food safety (Anonymous, 1998-I). Safety hazards may originate at any point in a production process, including receipt of raw materials, food handling, storage, packaging, and transportation. This concept is already being used across the food industry, especially in large food companies, and it is also very popular in the manufacture of MAP products. From the agricultural point of view, knowledge about the plant product and the details of the whole processing are basic tools necessary to begin to apply the HACCP method. To apply HACCP the following points must be known:

- Primary microorganisms and their sources
- Influence of temperature on pathogens
- Other types of contaminants from chemical or physical sources

HACCP simplifies hazard analysis by focusing on three types of contaminants: microbiological (pathogens), chemical (pesticides), and physical (any kind of foreign material) (Barron, 1996). In the next few years, when the rules and regulations of the World Trade Organization will be promulgated and implemented, *Codex Alimentarus* and HACCP applications will act as reference sources for many fields including MAP production. To facilitate the application of these rules there are also some guides such as *WHO Monographs on Selected Medicinal Plants* (Anonymous, 1999) and *Quality Control Methods for Medicinal Plant Materials* (Anonymous, 1998-II). These references provide general systematic information on the safety, efficiency, and quality control/quality assurance of widely used medicinal plants to facilitate their appropriate use.

Within the very detailed and multidisciplinary coverage of these parameters, the responsibility of agriculture requires clear definition. From the agricultural point of view, the main purpose is to produce the plant with high active ingredient content and to protect it with minimum losses. To verify whether this aim is achieved, the purity of MAP is determined and chemical assays are carried out.

Purity of MAP

To evaluate the purity of plant material sensory, macroscopic, and microscopic properties are determined. The macroscopic identity of MAP is based on shape, size, color, surface characteristics, texture, fracture characteristics, and appearance. However, since these characteristics are judged subjectively and substitutes or adulterants may closely resemble the genuine material, it is often necessary to substantiate the findings by microscopy or physicochemical analysis. By macroscopic examination it is also possible to determine the presence of foreign matter in whole or cut plant materials. Foreign matter is material consisting of any or all of the following:

- parts of the MAP or materials other than those named with the limits specified for the plant material concerned
- any organism, part or product of an organism, other than that named in the specification and description of the plant material concerned

- mineral admixtures not adhering to the medicinal plant materials, such as soil, stones, sand, and dust that must be removed before plant materials are processed (Anonymous, 1998-II)

MAP should be entirely free from *pesticide residues, arsenic* and *heavy metals, bacteria* and *mold, animal* and *radioactive contamination.*

Naturally grown or cultivated MAP are liable to contain *pesticide residues* that must be controlled according to the standard procedures of FAO, which includes the analytical methodology for the assessment of specific pesticide residues (Anonymous, 1997-II).

The contamination of medicinal plant materials with *arsenic* and *heavy metals* can be attributed to many causes, including environmental pollution and traces of pesticides. The amount of arsenic in the medicinal plant material is estimated by matching the depth of color with that of a standard stain (Anonymous, 1997-II). The quantity of Zn, Cu, and Mn could be in correlation with the herbicide treatments. The contents of Pb and Cd are the consequence of air pollution and fuel impurities, which were higher in flowers of *Matricaria chamomilla* L. grown near roads (Šovljanski et al., 1989).

Medicinal plant materials normally carry a great number of *bacteria* and *molds,* often originating in soil. While a large range of bacteria and fungi form the naturally occurring microflora of herbs, aerobic spore-forming bacteria also frequently predominate. Current practices of harvesting, handling, and production may cause additional contamination and microbial growth. The determination of *Escherichia coli* and molds may indicate the quality of production and harvesting practices (Anonymous, 1997-II). Methods for decontamination are restricted. For example, the use of ethylene oxide has been forbidden within countries of the European Union. Treatment with ionizing irradiation is also forbidden or requires a special registration procedure in some countries. In addition, the presence of Aflatoxins in plant material can be hazardous to health if absorbed even in very small amounts. They should therefore be determined after using a suitable clean-up procedure. Different limits for microorganisms are set according to the use of the material and the material itself (Anonymous, 2002). MAP should be completely free from toxins.

Dangerous *radioactive contamination* may be the consequence of a nuclear accident. The range of radionuclides that may be released into the environment as the result of a nuclear accident might include long-lived and short-lived fission products, actinides, and activation products. The amount of exposure to radiation depends on the intake of radionuclides and other variables such as age, metabolic kinetics, and the weight of the individual.

Biochemical and Sensory Characteristics of MAP

Within the framework of the biochemical characteristics of MAP are biological active ingredients, extractable matter, ash, bitterness value, hemolytic activity, swelling index, and foaming index, which are examined (Anonymous, 1998-II).

Biological active ingredients are chemical compounds of plant material that have a biological effect on other organisms. Biological effect comes about usually by physiological means or by secondary metabolites. Plants produce these compounds normally for their own ecological functions such as protecting against stress factors. Some MAP are particularly rich in some compounds that determine the final usage. The content and composition of the active ingredient, as well as the organoleptic characteristics are of crucial importance. The combination of active ingredients in herbal remedies produces synergetic effects that are more effective than that of each individual component.

The main factors determining the chemical composition of MAP are essential oils, carbohydrates, proteins, lipids, phenols, tannins, acids, vitamins, enzymes, color components, minerals, antimicrobials, sulfurs, resins, and terpenes (Akgül, 1993). The active ingredients are concentrated in the different vegetative parts of a plant such as the stigma, bud, flowers, fruit, fruit skin, seed, leaf, root, rhizome, and bulb. Many MAP contain most of the active ingredients given above but in different amounts depending on the plant species. The quality of active ingredients varies depending upon the ecological conditions, cultivation practices, harvest, and postharvest handling.

Good Agricultural Practices

Using inadequate input that has significant negative impacts on the environment and the consumer makes agricultural production unprofitable. Sustainable production systems and management of natural resources were discussed among all segments of society in the past decade for environment friendly and profitable farming. Sustainability of agriculture aims at socially viable and economically feasible systems where protecting the environment, and human and animal health are important considerations. Good agricultural practices (GAP) is a general terminology that defines the rules of sustainable agricultural production systems for governments, growers, suppliers, processors, traders, and consumers (Anonymous, 2003-III). GAP acts as a guide to unify national and international policies and methods on sustainable agriculture. GAP is not an official legislation; it contains voluntarily applied rules that could be developed by different nongovernmental organizations. Within the framework of GAP a set of principles for individual production systems is elaborated. Through these efforts it is aimed to draw all segments of society into the debate, into the action, and into the transition to a sustainable agriculture. The code of GAP does not mean that all goals are achieved and the work is completed. Owing to the dynamic nature of things the code should be developed further to include new requirements. MAP production on the other hand has some special characteristics that differ from commercial and horticultural crops. In this sector, there is overexploitation and illegal trade of wild species, habitat destruction, and inadequate harvest systems that have resulted in unsustainable usage of natural resources and threat to certain species. To provide some additional regulations for the cultivation and postharvest processing of MAP used as a raw material in food, feed, pharmacology, and cosmetic industry good agricultural practices and good wild crafting practice (GWP) for medicinal and aromatic plants have been developed by the European Grower Association (Anonymous, 2003-I). Understanding and accepting GAP for MAP cultivation without disturbing the wild habitat should be possible for the common benefit of all people.

In reality the rules of GAP are still not known to most growers. For example, the washing of herbs during the growing process is usually not practiced, which is important for herbs collected for their roots. During collection in wet conditions mud could be smeared on all roots and small rootlets. If the crop is not properly cleaned, more money could be gained because it is sold by weight. Sometimes unclean water causes additional bacterial contamination that has an impact on food safety. The other source of contamination is fertilizing with animal waste and drying on the ground. To determine bacterial contamination microbiological testing is required, which is carried out by the purchaser. There are technical solutions to these problems but it means more investment for the grower. Improper procedures, such as mixing contaminated crop with a clean batch, should be avoided as even the tiniest amount of bacteria could get into the stomach and breed there. Radioactivity is another important type of contamination that was intensively discussed after the Chernobyl incident. After the disaster, a very high quantity of radioactive contamination was determined in many species originating from different locations in Europe and Asia. Unfortunately, most of the contaminated herbs were not destroyed. They were mixed and blended with less contaminated crop to meet the specifications of companies. If these kinds of problems are not detected during purchasing, it would be very hard to control them in the finished product.

Mixing insect fragments, animal excrement, or rodent hairs to the crop is also possible in postharvest stages. The allowed quantity of these is strictly restricted by the row crop buyer or processor, who carries out not only sighting but also some floatation tests before purchasing. For testing, a sample is put into liquid where some of the material go to the bottom and some float up because of the difference in specific weight. Examination or skimming under a microscope makes it possible to trace this type of foreign material.

General problems of MAP cultivation that could be solved by implementing GAP are summarized as follows:

- Improper site selection from the agro-ecological point of view
- Using improper pesticides and herbicides
- Improper irrigation and fertilizing methods, timing and substances, using drainage water for irrigation
- Improper harvest time, method, and implementation resulting in quality and quantity losses
- Drying directly on the ground and in the sun, mixing various types of organic and inorganic foreign material to crop, quality reduction

- Using bacterial contaminated water from open pools for processing such as washing, distillation, etc.
- Insufficient hygienic measures in buildings, lack of knowledge on importance of personal hygiene
- Too much vehicular traffic in the farm and improper transport logistics from field to farm, quality reduction during transport because of fermentation (self-heating, etc.)
- Improper waste management, environmental pollution due to processing residues
- Drying facilities in farm buildings without any wire netting against birds, rodents, and domestic animals
- Problems resulting from reuse of unclean sacks, bags, or containers
- Insufficient storage measures, lack of knowledge of environment friendly fumigation substances, deterioration of crop in storage phase because of remoistening, pest attack, etc.

Growers have the important task and responsibility of solving these problems. First of all, they should adopt GAP and eliminate their poor practices for which the educational support of public extension services would be necessary. Growers should also be aware that if they persist with their wrong practices, it would not be possible to maintain consumer confidence in crop quality and safety. On the other hand, the adoption and implementation of GAP is the responsibility of not only the grower; a combined, coordinated, and integrated effort is required. This is achieved by sharing the responsibility among all segments in this sector. Thoroughly controlled production of MAP could be better promoted if unsustainable usage and the threat to certain species in spontaneous flora could be prevented. The question whether "wild collection from spontaneous flora" or "field cultivation" has been discussed for a long time among all segments of the MAP sector. Although producers prefer cultivated raw crop because of their reliability and continuity, the expectation concerning the solving of cultivation problems of many MAP species in the near future is not very realistic. Therefore, wild collecting should not be disregarded and efforts should be also made to regulate it. As such the rules for GWP of MAP could be an appropriate tool, until all demands of the market are covered by field production. The GWP of

MAP, which is developed by the European Grower Association, is intended to be voluntarily applied to the harvesting and processing of all MAP collected from wild resources. According to EUROPAM the reasons why it is more convenient to develop wild crafting production of medicinal and aromatic plants are:

- long growth periods for ripening and consequently for the harvest, in particular regarding trees and bushes;
- the plant cannot be normally cultivated for reasons not related to cultivation—symbiotic relationship with other plants;
- difficult to germinate, in finding the seeds, in transplanting, etc., which does not justify the time and the costs necessary for the use of the plant for cultivation purposes; and
- the quantity required of the plant is too exiguous to justify the economic costs of cultivation (Anonymous, 2003-I).

The guidelines of GAP and GWP are not the last word. They are open to modification and updated periodically. Each rule of both guidelines has been developed on the basis of practical experiences. In common practice, there are many cases confirming the importance of GAP and GWP. For example undesired mixing of similar plants could frequently happen during wild collecting. St. John's wort (*Hypericum perforatum*) is very often contaminated with *H. maculatum, H. crispum,* and *H. montanum*, some of which are free of the active ingredient Hyperforin that is essential in a variety of prescription products (Schneider, 2003). Likewise, Chamomile (*Chamomilla recutita*) can be easily confused with *Chamomilla inodorum* or *Chamomilla suaveolens*. In severe cases, such as mistaking *Petasis hybridus* (Butterbur) for *Tussilago farfara* (Coltsfoot), the resulting toxicity levels can be dangerous. Another example is the labeling that is very important from row crop harvesting to final product manufacturing. Most labels give little information about the plant material that was used to prepare the herbal remedies. It is well known that the quality and quantity of active compounds of each species could change depending on environmental and genetic variations. There are nine different species of *Echinacea,* for example (Simon cited in Anonymous, 2003-IV). Two of them are *Echinacea purpuea* and *Echinacea angustifolia* which are mixed in different quantities and used to remove symptoms of cold. Both plants produce compounds that have similar biological activity but in different amounts. If the name of the species is not indicated on the label at the production stage, the commercial value of the raw crop would decrease seriously, as it is not likely to be used for preparing some special herbal remedies.

Improving Quality

Global increase in trade of MAP leads not only to overharvesting and extinction of some species in habitat, but also to reduction in the quality and quantity of the raw crop because of picking at different vegetation stages and improper handling. Even when MAP species are safely cultivated, the quality cannot be ensured in cases of poor agricultural practice, such as mono cropping, or using inadequate inputs and the like. These are the most important negative impacts of the commercialization of MAP in terms of quality. In light of the changing demand dynamics of the growing market, it is clear that improving the quality of the raw material at the source of production should be the first aim for the healthy growth of the sector, because the quality of the raw material determines and indicates the quality of the finished product. To fulfill the demands of quality all steps of production, from cultivation to packaging, should be kept under control. The approach to promoting quality development in MAP production could be summarized as follows:

- International regulations should be followed and national rules for control of wild collecting from spontaneous flora should be defined and implemented.
- Collectors, growers, traders, manufacturers, and exporters should be informed on quality-related issues by governmental or non-governmental institutions.
- To promote traceability of the complete supply chain long-term relationships should be formed and maintained between manufacturer and reputable growers/suppliers.
- To meet the existing quality and traceability requirements and increased expectations, GAP and GWP should be implemented. There is also need for additional research on cultural practices such as soil preparation, propagation methods, sowing time, vegetation cycle, harvest and postharvest processing. Plant spe-

cific and praxis-oriented information for selected and tested species could be prepared by experts and distributed to growers.
- To implement good manufacturing practices (GMP) processors and manufacturers should be empowered. They should also provide infrastructure for ISO 9001/14001 and HACCP certification, which covers quality evaluation in areas of physical, chemical, and microbiological testing/analysis.
- Raw material should be sourced not through anonymous trading companies but through experts from the manufacturers themselves.
- Supply and demand development in world markets could change very rapidly and grower associations should keep a tab on it; if this facility does not exist locally, it should be established.
- Because of the increasing range in products, the industry prefers to invest in promoting the product rather than improving its production and quality. The industry could also support cultural production by using different tools such as financing of agricultural consultancy or investing proper inputs for growers.

Official data in producer countries do not usually reflect the real state of things, because in statistics data on exported MAP are usually placed under the section termed "others" where even the common name of the plant is not registered. On the other hand, exporters do not reveal the export data for the sake of privacy. Therefore, trade of MAP should be monitored and appropriately documented by governmental institutions.

Environmental Aspects

Environmental aspects could be evaluated from different points of view. But overexploitation of natural sources and environmental pollution are two important concerns related to MAP production.

Overexploitation

The growth in MAP demand in the world markets has led to an increased intensity and frequency of MAP harvesting from wild sources. Consequently, certain popularly traded species are overexploited and are now rare or extinct in the wild flora. This has resulted in the forced use of alternative species and a geographical shift of the cultivation process to previously unexploited areas (Diederichs et al., 2002). The number of threatened MAP species in the world is estimated to be between 4,000 and 10,000 in different reports (Vorhies, 2000; Schippmann et al., 2002).

The main result of overexploitation is partly or fully damaged flora that is not able to regenerate naturally. For example, intense and frequent bark harvest of species with a high-market demand often results in ring-barking of trees. Although it is commonly believed that most tree species completely regenerate their bark after it has been damaged, this is the exception, rather than a common response to serious bark damage (>10 percent of trunk bark removed below head height). In many reports, *Prunus Africana* (Pygeum) is given as an important example for bark harvesting (Cunningham and Mbenkum, 1995; Anonymous, 2004). The medicinal use of this species in Africa attracted consumers of the world markets, and the fast increasing demand has had a devastating effect on populations of this species. Debarked trees are very vulnerable to fungal or insect attack. Besides, regeneration of many threatened species could certainly come down because of unmanaged harvesting which means reduction of this species in flora. This also has negative impacts on food assurance of frugivorous animals. To protect all endangered species all efforts are being made from the perspective of nongovernmental institutions. Today many species are still considered critically endangered.

As discussed in the above sections cultivation of MAP is certainly necessary from the quality and quantity point of view. But, on the other hand it is also true that undesirable, destructive, and illegal wild harvesting is still maintained. To reduce the negative impacts of wild harvesting on nature, the sustainability concept is proposed by different experts (Schippmann et al., 2002). The basic tenet of sustainable harvest is nondestructive collection of MAP from spontaneous flora with the aspect of conserving the diversity of the ecosystem.

Environmental Pollution

Usually, it is believed that MAP-based products are safe and non-toxic. But, if plants are collected or cultivated in polluted environments where soil, water, and air contain high levels of toxic pollutants, the toxins can be translocated to humans and animals. Therefore, the possibility that toxic pollutants are deposited in the MAP cannot be disregarded. Pollution in MAP originates from the following sources:

- Uncontrolled use of pesticides, herbicides and fertilizers
- Accumulated toxic metals in soil and underground water and their transition to plants
- Previous and neighboring plants
- Toxic gases in the atmosphere emanating from nearby roads and industries and acid rains

The chemical residues in MAP products depend on many factors including application norm, duration between application and harvest, and the like. Therefore, a generalized accusative statement for the residues in MAP products cannot be made. For each species and application, specific research is required. For example, 2 weeks after azinphosmethyl treatment was applied at 1.1 kg/ha on *Mentha piperita* L. residue was detected in essential oil (Belanger, 1989). However, at the time of the harvest, 28 days later, there was no residue in the oil. Against that Cypermethrin applied at 0.2 l/ha for the same crop resulted in a residue level of 0.07 ppm at harvest. Except the scientific efforts all other sampling from commercial products show that about 4.3 percent of MAP samples contain high levels of residues (Zambo et al., 1988).

The reaction of MAP to chemicals applied in previous or neighboring plants has not been sufficiently investigated. The soil as an important pollution accumulator is primarily influenced by former applications. Normally, a new and clean production of MAP requires time to allow the soil to eliminate residues of the previous culture. Therefore, transition time is required for cleaning the soil of contaminants and the half time of conventional agricultural inputs can be too long as expected. The total removal of environmental pollutants within a short time period seems not very realistic. From the environmental point of view MAP could also be cultivated for phytoremediation, which is the use of plants to remediate contaminated soil and groundwater (Sutherson, 1999). Whether the harvested plant part could be used as a MAP product is another aspect. The transition mechanism of heavy metals from any plant part to a secondary product such as essential oil is not well known for many MAP species. According to some reports essential oil could be completely free of contaminants, although the harvested crop contains high level of pollutants. For example, *Mentha piperita* L. and *Mentha arvensis* L. could be easily grown in soils that are highly contaminated with Cd, Cu, Pb, Zn, and Cu (Zheljazkov et al., 1999). These species could be used for remediation purpose. The essential oil yield of these species cultivated in contaminated soil is lower than that of control, but no contaminant was found in oil. Review of previous researches shows that the knowledge about the status of toxic metals in MAP grown in polluted environments is not sufficient. For commercial use of remediated MAP species considerable research efforts are required. Nevertheless, to reduce problems related to environmental pollution good agricultural practices and good wild crafting practices for MAP could be generally referred to as important guides.

MECHANIZATION OF MAP PRODUCTION

Agricultural mechanization plays an important role in reaching profitable production. For each cultivated species of MAP a specific mechanization chain or specific implement for every step of production is not available. The actual situation for the mechanization of MAP could be presented as follows:

- There is a general lack of mechanization.
- Mechanization produced and used in developed countries is not achievable and profitable for small and medium enterprises with the exception of cooperatives and well-organized machine rings.
- Because of the limited market share of MAP in total agricultural production, the number of required specific machines in the world market does not allow for economically feasible manufacturing. The development costs of these machines are very high. Logically, the big manufacturers would not want to invest to produce only a few machines. On the other hand, small manufacturers can flexibly produce individual machines according to the specific requirements of growers.

- Because of the high diversity, different species, and used parts of plants, it is not possible to develop an universally applicable machine for all MAP.
- To solve their specific problems growers engage in designing and manufacturing specific machines, machines that are custom-built.
- Complete mechanization, as is prevalent for other agricultural crops, is not absolutely necessary. The term *humanization* is very important for MAP production where improvement of working conditions from the ergonomic point of view is intended. Increasing work efficiency and reducing the drudgery of using small tools and devices are important aspects for machine development.

Development of Procedures

Agricultural mechanization is not a problem related to only engineering. Success in procedure development and mechanization in rural area also depend much on politics, economics, and sociology. The isolated engineering solutions from real conditions usually remain as theoretical approaches that cannot contribute to solving the actual problems. Martinov et al. (1997) state that the following points should be considered for successful mechanization of special crops including MAP:

- Technology and mechanization must be relevant to social goals and applicable to the particular area.
- The solutions of developed countries are not directly applicable and are usually too expensive for the growers in developing countries.
- For identification of mechanization requirements and machinery selection the creative potential of local growers must be evaluated. It is expected that the growers ideas should contain the elements of social conditions and influences, as well as the understanding of local needs and manufacturing possibilities. The mechanization that is introduced should fulfill the social, cultural, employment, and production needs of the local growers.

- For mechanization self-reliant approaches based on indigenous manufacture with optimum use of local labor, production methods, and available materials should be adopted.
- To support the local manufacturing of the required machinery, governmental or public institutions, having staff with engineering skills, are of crucial importance. Engineering knowledge is necessary but not sufficient for realization of adequate technique and machinery. Therefore, the team for machinery development could include a cultural anthropologist and rural sociologist, in addition to engineers.
- Help of institutions or financial support from government in realizing prototypes of machinery should be obligatory and the know-how disseminated by publishing of obtained results.

To develop a machine for MAP the flow chart in Figure 1.2 could be followed. The activities on the left are carried out by growers, and the ones on the right by development institutions. The first step is identification of the need for the machine. In the second step, the activities of growers and institutions are carried out separately. In the frame of institutional activities engineers search for new ideas. They animate and stimulate the growers, who should express their ideas at this stage. It is expected that the described role of growers could sublimate all peculiarity of MAP production and local conditions. Engineering knowledge should help to overcome potential engineering mistakes and possible disorientation from growers. Growers who are involved in production problems are a great potential for contributing new ideas. The next two steps are evaluating of ideas and selecting one or two of them as prototype. These steps have already been practiced in developed countries for a long time. Engineering articulation of a selected idea for prototype is the responsibility of professionals from institutions who work on the basic principles of the machine design, which should be inexpensive, made of locally available raw materials, using standard elements, and reflect the technique of local manufacturers. Engineers should accept technical and economical evaluation of design used in developed countries. Making a prototype could be realized in the institution, by local manufacturer, or even in the grower's workshop. The production costs could be covered by the grower who is the owner of the prototype. A part of the cost could be

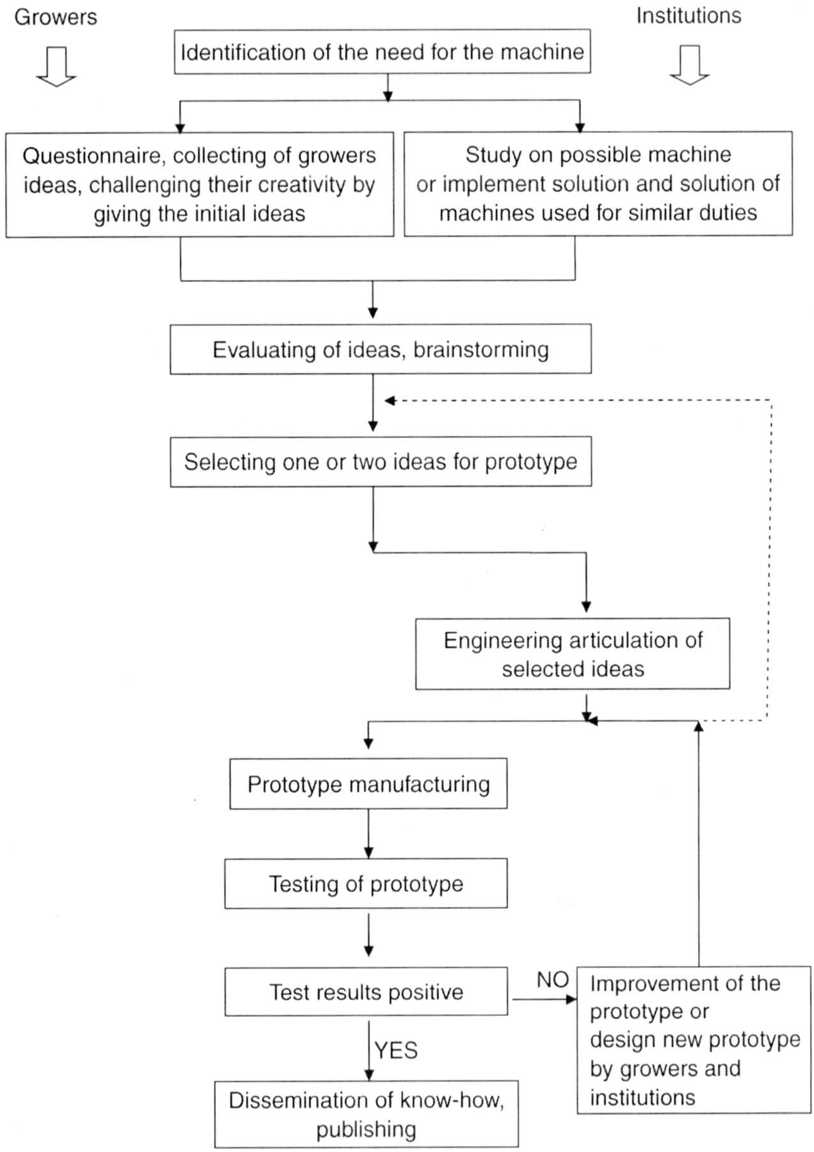

FIGURE 1.2. Flow chart to develop a machine for MAP. *Source:* Martinov et al., 1997.

supported by governmental budget. Institutional services should be financed by public sources responsible for agricultural development.

Testing and improving of the prototype and evaluation of the results achieved is a joint task for both the grower and institution. If the test results are successful, they should be disseminated by each possible tool. At the end of this process, the ownership of the experience reached and the knowledge do not belong to one person or group. The participation of society obliges the grower and institution to publish the new technique and mechanization for producing special culture such as MAP. The proposed method has been applied and improved by the development of some new machines, primarily for MAP (Martinov and Müller, 1990; Martinov et al., 1992; Soysal and Öztekin, 2001).

REFERENCES

Accascina G. 2000. Information technology and poverty alleviation. SD dimensions of FAO. Available from: http://www.fao.org/sd/CDdirect/CDre0055h.htm. Accessed 2003 September 12.

Akgül A. 1993. *Spice Science & Technology* (in Turkish). Ankara-Turkey: Food Technology Association, 15, 450 pp.

[Anonymous] 1992. Resource guide to growing and using herbs. Available from: http://www.nal.usda.gov/afsic/AFSIC_pubs/srb93-01.htm. Accessed 2003 November 22.

[Anonymous] 1997-I. Plant talks: 100 plant facts for campaigning conservationists: plants in traditional and herbal medicine. Available from: http://www.plant-talk.org/Pages/Pfacts10.html. Accessed 2003 June 17.

[Anonymous] 1997-II. Codex maximum limits for pesticide residues. 22nd Session (June 1997) of Codex Alimentarius Commission (available in electronic format from FAO. Available from: http://apps.fao.org/CodexSystem/pestdes/pest_ref/pest-e.htm. Accessed 2003 August 14.

[Anonymous] 1998-I. Guidance on regulatory assessment of HACCP—The report of a Joint FAO/WHO Consultation on the Role of Government Agencies in Assessing HACCP (WHO/FSF/FOS/98.5). Available from: http://www.who.int/fsf/REP983A.html. Accessed 2003 August 9.

[Anonymous] 1998-II. Quality control methods for medicinal plant materials. Available from: http://www.who.int/medicines/library/trm/medicinalplants/qualitycontrolmeth.pdf. Accessed 2004 September 22.

[Anonymous] 1999. *WHO Monographs on Selected Medicinal Plants*. 3 volumes. Geneva: WHO.

[Anonymous] 2002. *Europaeisches Arzneibuch. Amtliche deutsche Ausgabe*. 4. Ausgabe. Stuttgat/Eschborn: Deutscher Apotheker Verlag/Govi Verlag.

[Anonymous] 2003-I. Good Agricultural Practices (GAP) and Good Wild Crafting Practices (GWP) of Medicinal and Aromatic Plants. The European Herb Growers Association (EUROPAM), GAP/GWP Subcommittee, EUROPAM Working Copy Nr. 3. Available from: http://www.europam.net/Working%20 documents .htm. Accessed 2003 September 12.

[Anonymous] 2003-II. What is Codex? Available from: http://www.who.int/fsf/ Codex/whatiscodex.htm. Accessed 2003 August 9.

[Anonymous] 2003-III. Good agricultural practices. FAO draft from October 2001. Available from: http://www.fao.org/prods/GAP/FAO-GAP.pdf. Accessed 2003 July 10.

[Anonymous] 2003-IV. Quality of herbal remedies is often guesswork. Available from: http://oci.mcw.edu/article/954385222.html. Accessed 2003 July 15.

[Anonymous] 2004. Botanic gardens and the sustainable use of plants. Available from: http://www.bgci.org.uk/plants/sustainable_use.html. Accessed 2003 June 17.

Barron FH. 1996. *HACCP and Microbreweries: Practical Guidelines of Food Safety for Microbreweries, Brewpubs and the Beer Industry.* Clemson, SC: Clemson University Cooperative Extension Service, EC 691,14 pp.

Baser KHC. 1997. Industrial utilization of medicinal and aromatic plants. Proceedings of Second World Congress on Medicinal and Aromatic Plants for Human Welfare, 10-15 November 1997, Mendoza, Argentina. Martino V, Bandoni A, Blaak G, Capelle N, editors. *Acta Hort;* 503: 21-29.

Belanger A. 1989. Residues of Azinphosmethyl, Cypermethrin, Benomyl and Chlorotalonil in Monarda and Peppermint Oil. *ACTA Horticulture;* 249: 67-73.

Bomme U. 1992. Medicinal Plant Production (In German). *Bodenkultur und Pflanzenbau;* 2/92: 3-12.

Ceylan A. 1987. *Medical Plants II* (in Turkish). First Edition. Izmir-Turkey: Agricultural Faculty of Aegean University, 188 pp.

Ceylan A. 1995. *Medical Plants I* (in Turkish). Third Edition. Izmir-Turkey: Agricultural Faculty of Aegean University, 140 pp.

Comer M, Debry E. 1996. A partnership: biotechnology, biopharmaceuticals and biodiversity. In: *Biodiversity, Science & Development,* Di Castri F, Younnes T. editors. Oxford: CAB International, 488-499.

Cunningham AB, Mbenkum FT. 1995. Sustainability of harvesting Prunus africana bark. Medicinal Plant in International Trade. People and Plants Working Paper-2, May 1995. Available from: http://www.rbgkew.org.uk/peopleplants/wp/wp2/ content.htm. Accessed 2003 July 23.

Davis JM. 2001. Production of medicinal herbs. Available from: http://fletcher.ces .state.nc.us/staff/jmdavis/patalk.html. Accessed 2003 August 15.

Diederichs N, Geldenhuys C, Mitchell D. 2002. The first legal harvesters of protected medicinal plants in South Africa. Available from: http://www.science inafrica.co.za/2002/november/bark.htm. Accessed 2004 November 15.

Heyvood V. 1999. Plant resources and their diversity in the Near East. Medicinal, aromatic and culinary plants in the Near East. Proceedings of the International Expert Meeting organized by FAO:19-21 May 1997. Cairo, Egypt. Available from: http://www.fao.org/documents/show_cdr.asp?url_file=/docrep/X5402e/ x5402e03.htm. Accessed 2004 November 24.

Koyuncu M. 1995. Medicinal and aromatic plants in Turkey. Available from: http://www.fao.org/docrep/X5402e/X5402e16.htm. Accessed 2003 June 15.

Kuipers SE. 1995. Trade in medicinal plants. In: Medicinal plants for conservation and health care. Available from: http://www.fao.org/documents/show_cdr.asp?url_file=/docrep/W7261e/W7261e08.htm. Accessed 2004 November 23.

Lambert J, Srivastava J, Vietmeyer N. 1997. *Medicinal Plants: Rescuing a Global Heritage.* World Bank Technical Paper: 355, Washington, 65 pp.

Lange D. 1998. Europe's medicinal and aromatic plants: their use, trade and conservation. A TRAFFIC Species in Danger Report. Available from: http://www.traffic.org/plants/executive_summary.html. Accessed 2003 September 22.

Lewington A. 1993. *A Review of the Importation of Medicinal Plants and Plant Extracts into Europe: TRAFFIC International.* Cambridge, UK: Traffic International. 37 pp.

Martinov M, Müller J. 1990. Production of peppermint in Yugoslavia (In German). *HGK Mitteilungen;* 33(4): 35-36.

Martinov M, Tešič M, Müller J. 1992. Harvest machine for Camomile (In German). *Landtechnik;* 47(10): 505-507.

Martinov M, Lammers PS, Tešič M. 1997. Involving growers in development of mechanization for special crops. *AMA;* 28(2): 65-68.

Munyua H. 2000. Information and communication technologies for rural development and food security: Lessons from field experiences in developing countries. SD dimensions of FAO. Available from: http://www.fao.org/sd/CDdirect/CDre0055b.htm. Accessed 2003 September 22.

Nair MNB, Ganapathi N. 1998. Medicinal plants: Cure for the 21st century: Biodiversity, conservation and utilization of medicinal plants. Seminar, 1998 October 15-16. Available from: https://www.vedamsbooks.com/no14512.htm. Accessed 2004 November 23.

Nickel W, Sennhauser E. 2001. Medicinal plants: local heritage with global importance. South Asia Brief. Available from: http://www.worldbank.org. Accessed 2003 September 12.

Schippmann U, Leaman DJ, Cunningham AB. 2002. Impact of cultivation and gathering of medicinal plants on biodiversity: global trends and issues. In: Biodiversity and the Ecosystem Approach in Agriculture, Forestry and Fisheries. Satellite event on the occasion of the Ninth Regular Session of the Commission on Genetic Resources for Food and Agriculture. Rome, 12-13 October 2002. Available from: http://www.fao.org/DOCREP/005/AA010E/AA010e00.htm. Accessed 2003 July 28.

Schneider M. 2003. Quality assurance from the ground up: sourcing raw material for the herbal marketplace. Available from: http://www.linnea-worldwide.com/PDFS/QualAssurGrUpReport.pdf. Accessed 2003 September 21.

Šovljanski R, Lazic S, Kišgeci J, Obradovic S, Macko V. 1989. Heavy metals contents in medicinal and spice plants treated with pesticide during the vegetation. *Acta Hort;* 249: 51-60.

Soysal Y, Öztekin S. 2001. Technical and economic performance of the tray dryer for medicinal and aromatic plants. *Journal of Agricultural Engineering Research;* 79(1): 73-79.

Srivastava J, Lambert J, Vietmayer N, editors. 1996. Medicinal plants: an expanding role in development. World Bank Technical Paper. No: 320. Available from: http://www.wds.worldbank.org/servlet/WDSContentServer/WDSP/IB/1996/04/01/000009265_3961219094248/Rendered/ PDF/multi_page.pdf. Accessed 2004 November 22.

Sutherson SS. 1999. *Remediation Engineering: Design Concepts.* New York. CRC. 351 pp.

Vorhies F. 2000. The global dimension of threatened medicinal plants from a conservation point of view. In: *Medicinal Utilization of Wild Species: Challenge for Man and Nature in the New Millennium.* Honnef S. and Melisch R. Editors. WWF Germany/TRAFFIC Europe. Hannover, Germany, p. 26-29.

Zambo I, Tetenyi P, Bernath J. 1988. Experiences on pesticide residues of medicinal plants in Hungary. *Acta Hort;* 249: 97-104.

Zheljazkov VD, Jeliazkova EA, Craker LE, Yankov B, Georgieva T, Kolev T, Kovatcheva N, Stanev S, Margina A. 1999. Heavy metal uptake by mint. *Acta Horticulturae* 500: II WOCMAP Congress Medicinal and Aromatic Plants. Part I: 111-118.

Chapter 2

Harvesting

Milan Martinov
Miodrag Konstantinovic

INTRODUCTION

MAP harvesting—harvesting of medicinal and aromatic plants—is a term that implies gathering of plant part or parts containing active ingredients. The term also includes preharvesting operations. One example of a preharvesting operation is the removal of aerial plant parts before digging out the roots or bulbs. The other example is the procedure of desiccation, a method that used to be based on chemical, and now only on mechanical, treatment.

The general requirements for successful MAP harvesting are more or less the same as for other crops. The plants, or their parts, should be harvested at the time of maximum yield and quality of material. A specific characteristic of MAP is that only one or some parts of the plant are active ingredient holders. That means that it would be more beneficial to harvest only those parts, and not the whole plant. Those parts of the plant without active ingredients, or the parts with negligible active ingredient content, are undesirable in the harvested material and are even declared as impurities.

MAP harvesting is termed successful if the harvesting procedure is realized in such a manner as to provide the best possible quality of harvested material, if damages of yielding parts are reduced as far as possible, and if it is ensured that there are no/or there is only minimal content of other plant parts and impurities.

Medicinal and Aromatic Crops

doi:10.1300/5895_02

In most cases, the period for a successful harvest is very short. Typically, it lasts one to two weeks. Therefore, good planning and organization are necessary with optimum utilization of the available machinery and labor capacity. Sometimes unfavorable weather conditions (e.g., rainfall, wind, and high air humidity) have a negative impact on the harvesting process. Notwithstanding, the harvest should be planned, organized, and realized to be completed in as short a period of time as possible. Otherwise, it might result in quality reduction or yield losses.

The harvesting can be classified according to different criteria. One of the criteria is the origin of the plant. The classification according to this criterion is as follows:

1. Harvesting of spontaneous flora (wild crafting)
2. Harvesting of cultivated plants

The harvesting of spontaneous flora, wild crafting, is still widely practiced in developing countries.

The second classification of harvesting is based on the procedure followed:

1. Manual
2. Semi-mechanized
3. Mechanized

Manual harvesting, if performed by skilled personnel, yields superior quality produce. In some cases, this is the only available procedure. This is also a labor-intensive procedure, and it cannot be profitable in developed and most of the developing countries, even if the labor costs are low. The demerit of this procedure is the difference in the quality of harvesting when performed by different persons.

Mechanized procedure comprises the use of machines with a significantly lower input of human labor. The quality of the harvested material could be lower or higher depending on the plant that is being harvested, the machinery used, and other influences. Well-planned drying, processing, and finalizing of a product are usually enough to eliminate this demerit. Mechanization can be of crucial importance to the financial success of MAP production, and it can contribute to the needed uniformity of products. The creation of new MAP varieties and the development of new techniques should be in the direction of enabling more successful mechanized harvesting. A synergetic approach in that direction is necessary.

The mechanization know-how available on the market is very expensive for MAP producers, and mechanization development on one's own is in most cases even more expensive. A good compromise is the development and utilization of auxiliary devices/tools that can make manual harvesting easier, more efficient, and work more ergonomic. This is the so-called humanization of work or semi-mechanization as it is otherwise known. It has positive economic and social effects, because it still enables employment of local labor, and at the same time reduces or eliminates drudgery. The development of full-scale mechanization for the developing countries can sometimes be even counterproductive. It might eliminate the demand for labor and might even enable the developed countries to produce the same product at competitive prices. Although this statement does not reflect the global progress, it considers a very important social component of the sensitive world rural development policy.

MANUAL AND SEMIMECHANIZED HARVESTING

The harvesting of spontaneous flora is nowadays performed manually or, occasionally, by semi-mechanized means. Figure 2.1 shows the picking of wild pepper in Africa.

It would be difficult to find a mechanized alternative for this type of harvesting. The only other possibility would be the use of equipment to enable a more comfortable position for the pickers, for example, a ladder.

In some countries manual harvesting of certain MAP is done extensively, with the use of certain appliances. One example is chamomile picking, where tools like a shovel with a comb or a similar tool with an additional rod are used (Figure 2.2).

The flowers are stripped by swinging the tool through the chamomile canopy, whereby, depending on the plants' height, shorter or longer stems are harvested. After every swing the picker breaks off the longer stems of the plants by hand. This operation results in better quality of yield but prolongs harvesting. With these tools some 400 man hours of skilled pickers are needed for harvesting one hectare.

FIGURE 2.1. Manual harvesting of wild pepper in Africa. *Source:* Reprinted with permission from Müller, 2000.

In Argentina and other countries a larger comb mounted on wheels has been developed and used, Figure 2.3. Two pickers can accelerate this device toward an unpicked canopy, with the comb angled in such a manner as to create the same effect as by two devices.

The same or similar devices can be used for picking of other flowers, such as marigold and pyrethrum. The problem is that the picking has to be selective, of only mature flowers of MAP, for example, marigold. The effectiveness of the picking and the length of the stems are strongly dependent on plant conditions. In the case of chamomile, the effectivity changes during the day with the change of air temperature and humidity. In the early morning and during the night the part of the stem under the flowers is more brittle and harvesting results are better. This problem could be theoretically overcome by breeding varieties that have weaker stems beneath the flowers.

FIGURE 2.2. Manual chamomile harvesting with simple devices. (a) simple shovel with nails forming a comb, (b) a similar type of device with a rod extension.

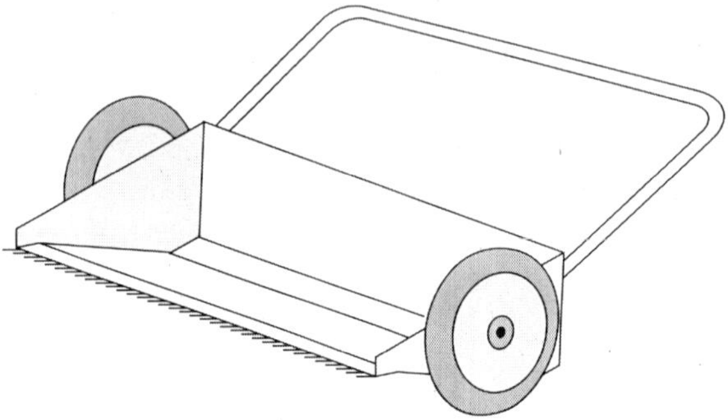

FIGURE 2.3. "On wheel" comb for chamomile harvesting.

The harvesting of herbaceous MAP by mowing with scythes or sickles is a common procedure for small-plot cultivation. The mowed material is collected, left on the stubble in swaths, or spread for on-field drying.

Cutting herbaceous plants using a knife, sickle, or other tools and tying them up into bundles is a traditional procedure in almost the whole of the Mediterranean region, Figure 2.4. This is used for the production of both fresh and dry market-ready MAP bundles.

MAP in the form of bundles, fresh or dry, could be successful on the market in the future, especially for exclusive hotels or special shops. The development of auxiliary devices for the harvesting and bundling of MAP is feasible, but it is a very cost-intensive procedure. The general idea of such a device is shown in Figure 2.5. The proposed device should have three different drives: for cutter bar, bands, and binding unit, which makes the structure complicated.

The posture of pickers during some traditional harvesting procedures is very inconvenient. Typical examples are shown in Figure 2.6, during the picking of saffron by twisting the flowers, which is a very uncomfortable posture to be in for a long time.

In such cases, semi-mechanization or humanization of work can be achieved by the use of a device that enables a more ergonomic position during harvesting. The idea for such a device or vehicle is shown in Figure 2.7. The vehicle can be driven by a worker and powered by an internal combustion engine or electro motor. The position of the picker should be adjustable, first according to anthropometrics,

FIGURE 2.4. Manual cutting and forming of bundles of herbaceous MAP in Mediterranean region.

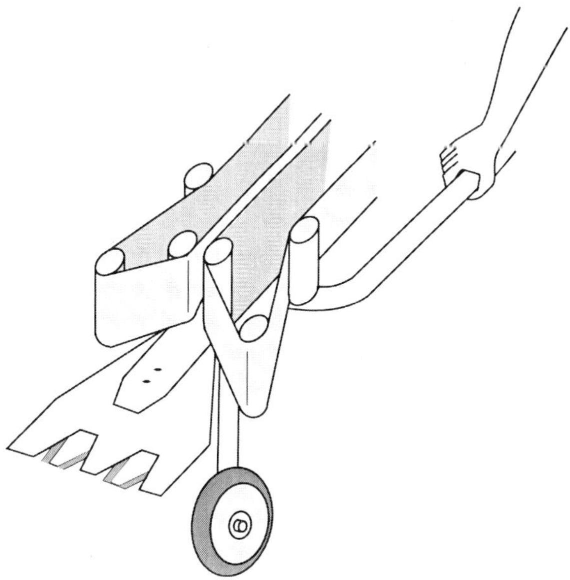

FIGURE 2.5. Device for MAP harvesting in a form of bundles—idea sketch.

FIGURE 2.6. Twisting of saffron flowers.

and second to enable the changing of position during the day. This solution has not only functional and human aspects but social ones too. It should attract young people to do this job and preserve the tradition of MAP production in the region.

Improvement of worker ergonomics has been already realized for some other crops. Figure 2.8 shows the harvesting of asparagus.

There are no serially produced devices specially aimed at wild crafting. The market is nowadays well supplied with different garden equipment powered by internal combustion engines at reasonable prices. Based on this, diverse mechanized devices for spontaneous flora harvesting can be developed. Some of the tools available are presented in Figures 2.9a and 2.9b. The development of special tools is also possible. The example of a stripper is presented in Figure 2.9c.

Contemporary flexible shafts can enable powering of different picking devices, and in most cases pneumatic transport of harvested material could be used. Ventilators generating airflow for this transport can be powered by the same shoulder-hitch engine, Figure 2.10. Of course, the problem would be to carry the harvested material. That can be overcome by using a small trailer pulled by the picker, which could hold the box for the harvested material, engine, and ventilator.

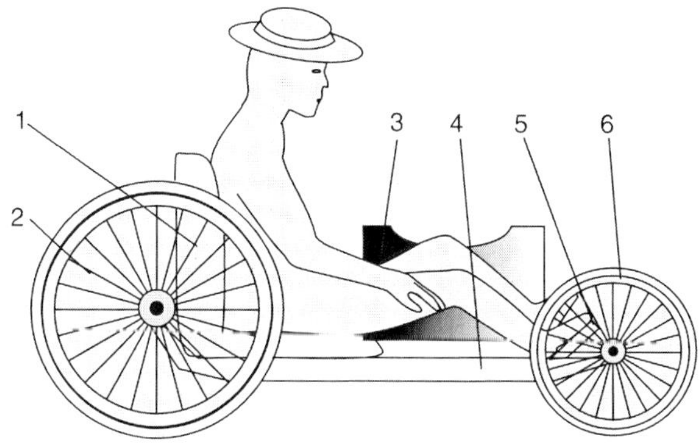

FIGURE 2.7. Vehicle for ergonomic positioning of pickers. 1–rear wheel, 2–adjustable seat, 3–basket for material–both sides, 4–frame, 5–pedals, 6–front wheel.

FIGURE 2.8. Vehicle used for asparagus harvesting, driven electrically.

The concept shown in Figure 2.10 should be adjusted for particular conditions, terrain configuration, and the local approach to harvesting and to the pickers' expectations. The ideas given are meant to be an inspiration for producers' innovations, in the manner described in Chapter 1.

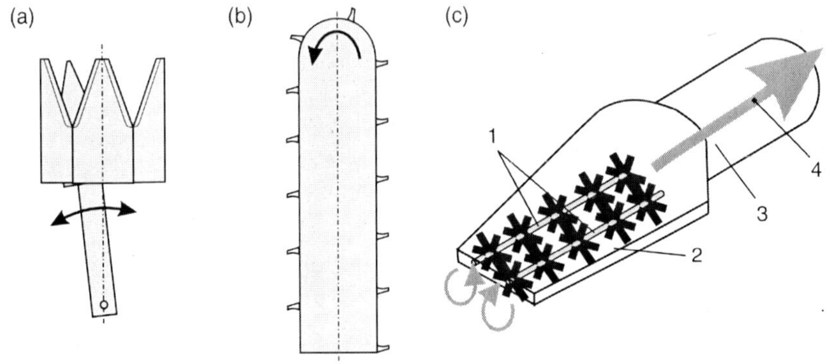

FIGURE 2.9. Potential tools for spontaneous flora harvesting. (a) cutter bar section, (b) sword cutter, two examples of mass products, (c) special powered stripper that can be used for some crops; 1–shafts with fingers, 2–housing, 3–tube, 4–air flow–transport. *Source:* Martinov et al., 1999.

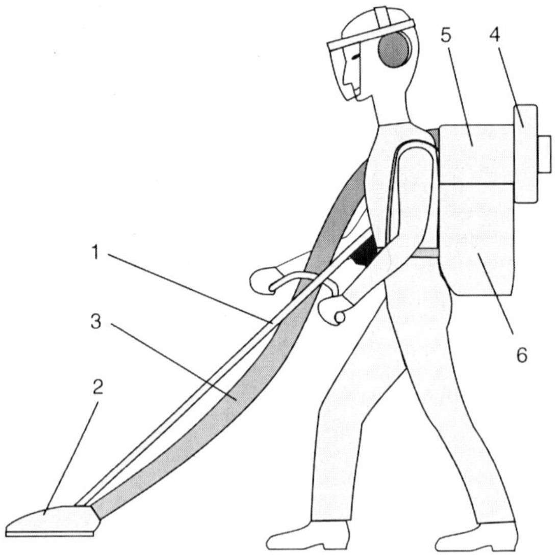

FIGURE 2.10. Idea for engine-powered harvesting equipment for spontaneous flora. 1–rod, 2–picking tool, 3–tube, 4–ventilator, 5–drive, 6–box. *Source:* Martinov et al., 1999.

MECHANIZED HARVESTING

Mechanized harvesting is one important prerequisite to achieve profitable production of MAP. Mechanization of harvesting is prevalent for cultivated plants. Agrotechnology, the production chain, and even varieties should be adjusted to the possibilities and needs of mechanized harvesting. In some cases, mechanized harvesting results in inferior quality in comparison with manual harvesting, but it is usually uniform, which enables more efficient and less labor-intensive harvesting and postharvest processing.

There is a wide range of machines used for harvesting. The machines and procedures that are used in small farms are different in comparison with those used in big farms. The similarity is in developed and developing countries. (A very detailed review of the literature and harvesting practice of medicinal and aromatic plants, for example of Germany, a developed country, is given by Zimmer and Müller, 2003.) Generally, the idea is to adapt machines for mass crops or to use some components of those machines. There are also examples of specially developed machines for some crops.

One of the most significant classifications of mechanized harvesting is regarding the plant parts that have to be harvested. Roughly, this classification is as follows:

- Roots and bulbs harvesting
- Herbs, leaves, and stalks harvesting
- Flowers or flowers' parts harvesting
- Fruits and grains harvesting
- Special types of harvesting

Root and Bulb Harvesting

Root harvesting is the procedure of digging–lifting underground plant parts and separation of soil and other impurities, as much as possible. Harvesting also includes, for some plants, the cutting and removal of aboveground parts. Diverse facilities can be used for this operation, and the schema of a typical one is shown in Figure 2.11.

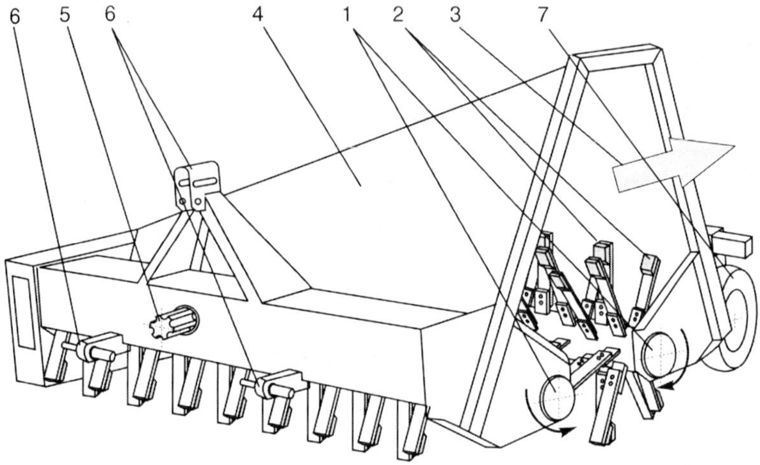

FIGURE 2.11. Machine for aboveground removal of plant parts. 1–High RPM shafts, 2–textile/rubber whips with metal endings, 3–direction of generated air flow, 4–housing, 5–connection to tractor PTO, 6–tractor fastening bolts, 7–ground wheels.

Root and bulb harvesting machines consist of two major parts: diggers and soil separators. The requirements for this procedure are as follows:

- To shear and dig the soil furrow deep and wide enough to include all yielding plant parts.
- Gently lifting of soil slice without damaging the parts of roots and bulbs.
- Effective separation of soil and other impurities by different harvesting procedures.

For some plants it is not so easy to fulfill specific requirements due to the shape of the roots and bulbs and aboveground architecture, Figure 2.12. Some of the MAP roots can be harvested by using the potato harvester–digger (Bomme, 1976).

One example of a special digger is a horseradish harvester, which harvests roots and bulbs with special configurations, Figure 2.12. Some special elements have been developed to enable effective dig-

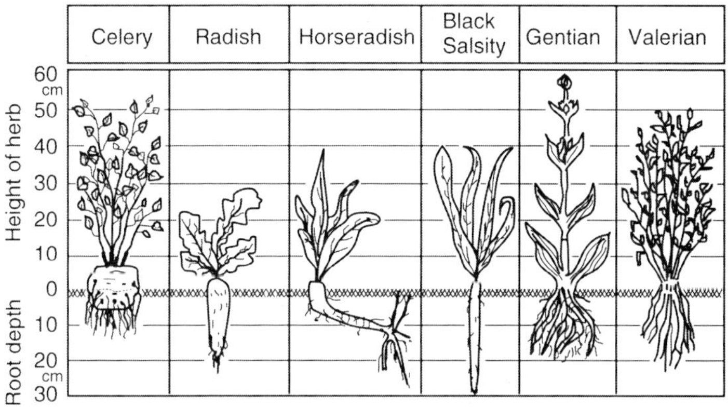

FIGURE 2.12. Root and bulb shape and aboveground architecture of some MAP roots.

ging of horseradish. A scheme of one of the developed elements is shown in Figure 2.13.

Two harvesters based on this operating principle are shown in Figure 2.14. Both machines have oscillating parts designed to reduce force. That enables more effective soil separation under different harvesting conditions (Sauter et al., 2002). The use of vibrators has or causes a negative impact to the tractor. The machines are operational to depths of up to 50 cm. For smaller farms a subsoiler type of row digger can be used. Depending on weather and soil conditions some additional sieving for soil separation may be needed.

For root and bulb digging the common moldboard ploughs can be used, with the soil sieving done afterwards. In that sense, the potato or onion digging machines are much more adequate. An example of a potato digger with two vertical counteroscillating units for MAP harvesting is shown in Figure 2.15.

Another possible adaptation of the conventional potato–onion harvesting machine is shown in Figure 2.16. The original potato-lifting share has been removed and a special fork-shaped tool for digging of MAP roots mounted (white ellipse). The belt chain elevator shuttled by the actuator intensifies soil separation. Such a machine design is acceptable for small and medium producers in developing countries.

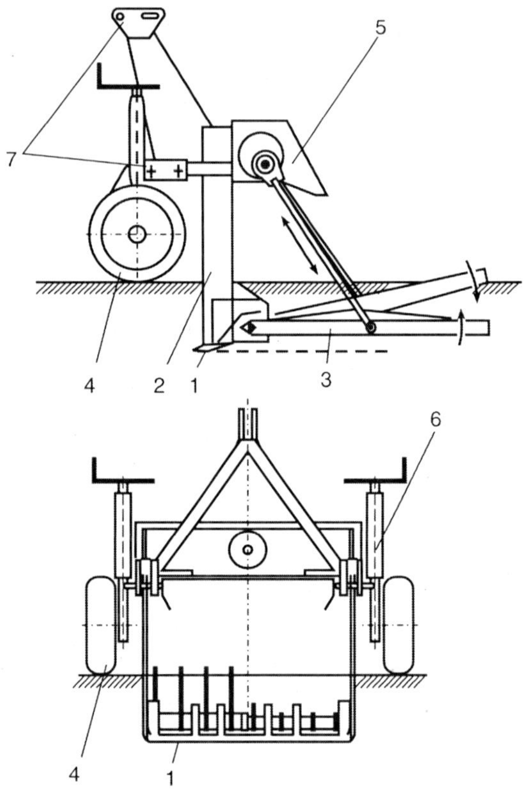

FIGURE 2.13. Horseradish digger. 1–lifting share, 2–frame, 3–counter phase oscillating soil removal grates, 4–supporting wheels, 5–tractor PTO driven eccentric, 6–wheels' adjustment, 7–tractor fastening bolts. *Source:* Schurig and Rödel, 1990. Reprinted with permission.

Valerian harvesting with an oscillating potato digger is presented in Figure 2.17. This is an example of machines originally designed for mass crops being used for harvesting of MAP.

Leaf and Stalk Harvesting

There are diverse possibilities to harvest herbs, leaves, and stalks. A typical procedure is to harvest the whole plant, for example, the

(a)

(b)

FIGURE 2.14. Specially developed horseradish diggers. (a) with two vibrating sections, first with share, (b) with three vibrating sections, stable share.

FIGURE 2.15. Harvesting of MAP roots with two vertical counteroscillating units of a potato harvester.

FIGURE 2.16. Adaptation of the conventional potato–onion harvesting machine for MAP roots, two pairs of additional digging tools. *Source:* Reprinted with permission from Müller, 2000.

common balm, peppermint, thyme, parsley, and the like. For successful harvesting the following criteria have to be fulfilled:

- Smooth cutting of stalks, without longitudinal split, or other destructions of harvested material or stubbles
- Cutting–lifting of all desirable material, including low-positioned plant limbs

- Gentle treatment of harvested material, for example low impact of reel support arm
- Gentle conveying of harvested material to the transportation trailer or machine hopper with minimum friction applied
- Minimal crop–stubble damage, primarily by tractor and machine tires

Different types of widely used grass mowers can be utilized for MAP harvesting, but most of them do not fulfill the defined specifications. Most of the grass or alfalfa mowers make swathes of the harvested material. This procedure is less convenient for MAP harvesting, due to the additional contamination of material with soil, but it is still carried out, especially if complete or partly natural drying is practiced. Conventional reciprocating cutter bars could be used, but according to the first requirement, the double knife type is more suitable. Rotary or disc mowers are suitable only for some MAP.

Conventional reels, used by universal combine harvesters, are too aggressive for some sensitive MAP. If there are possibilities of complex adjustment of the reel—position, RPM, type of fingers—the MAP harvesting demands could be fulfilled easily. By using additional devices, in front of the cutter bar, low-positioned limbs could also be harvested, but good leveling of the field is a prerequisite.

FIGURE 2.17. Valerian harvesting with potato harvester—oscillating type.

Problems occur when harvesting extremely low crops such as thyme. Slow speed, sharp cutter bar knifes, and special adjustment of the reel is needed. An even greater problem is the harvesting of young plants with undeveloped roots.

Most of the fresh plants are very sensitive and therefore the conveyance of these needs to be gentle. The most conventional mowers have an auger as a part of conveying system. It causes intensive rubbing, resulting in damages to active ingredient holders in most MAP. Typical grain conveyors of combine harvesters press the material and pull it over a sheet metal bottom. This causes even higher damages to herbaceous plants. The best solution is a band conveyor with free layers of transported material and no additional compression.

The most herbaceous of MAP are perennials, with few annual harvests. Harvest machinery always causes damage to plants and stubbles, as well as soil compression. This should be taken into consideration while planning the harvesting procedure, including transportation, and different possibilities should be evaluated to reduce damages of plants and soil compression. The tire width of tractors, machines, and trailers should be adjusted to fit the space between the rows of the crops.

The management of harvest should be evaluated as well. Quick transport of the harvested material to the dryer or processing facility is of great importance. The use of harvesting machines for transport is acceptable only for short distances, up to 5 km.

Most of the specified requests for a good MAP harvester have been considered and realized by the development of a self-propelled harvester Hege (Bomme et al., 2000). Figure 2.18 shows parts of this machine incorporated to improve its function in MAP harvesting.

Figure 2.19 shows the effects of the "torpedo" limbs' lifters by mowing of low limb level crops (b), compared with conventional cutter bar effects (a). Figure 2.19 shows the view of the cutter bar with "torpedoes" and reel with one type of free hanging brush for gentle positioning.

Reduced reloading capability to the transport trailer could be considered as a demerit of the machine shown in Figure 2.20. One solution, shown in Figure 2.21, with hydraulic lifting of the rear side of the hopper is designed to avoid reloading of the plant material. Un-

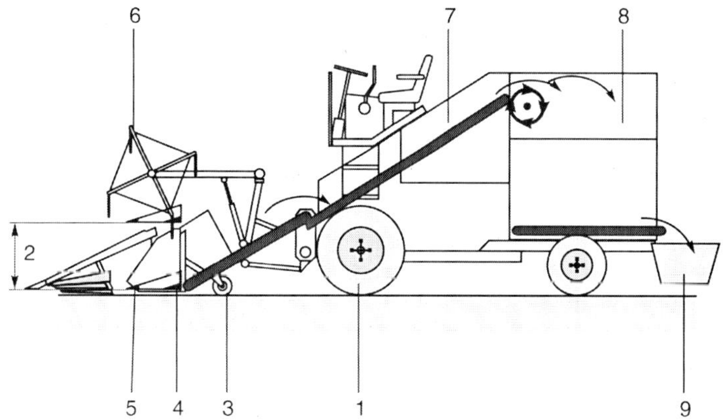

FIGURE 2.18. Special machine for harvesting of herbaceous MAP. 1–wheels width adjustable track gage, 1.25 m, 1.50 m, 1.60 m, and 1.90 m, 2–adjustable cutter bar height 50-800 mm, 3–cutter bar guide wheel, 4–double knife cutter bar, 5–adjustable reel with different types of free hanging brushes, 6–"torpedo" limb lifter, adjustable for every row, 7–belt conveyors, 8–hopper with discharging band, 9–hopper for small samples. *Source:* Bomme et al., 2000. Reprinted with permission.

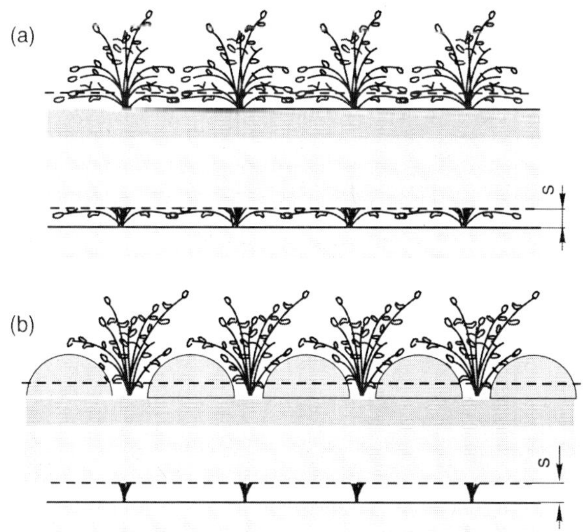

FIGURE 2.19. Low limb level crops mowing. (a) conventional mower, (b) mower with "torpedoes."

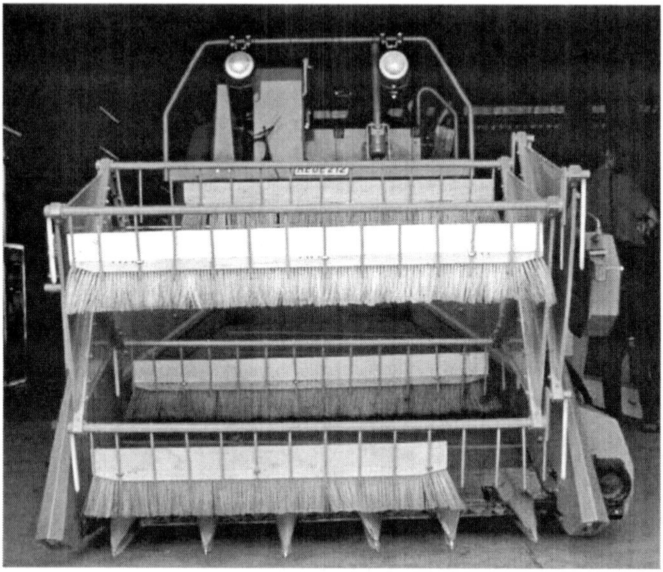

FIGURE 2.20. Herbaceous MAP harvester with double knife cutter bar, special reel and "torpedoes."

FIGURE 2.21. Herbaceous MAP harvester with hydraulic lifter for better discharge of material.

fortunately, this is not an ideal solution, due to the low discharging height.

The one-person operating machine that is described is very suitable for farmers in developing countries, but the high price is a decisive obstacle for its purchase. However, this is a good example of development, which has fulfilled almost all the demands for the successful harvesting of herbaceous MAP.

Some companies from the group of large-scale industries, made some attempts to step into the market for MAP producers. An example is the introduction of a self-propelled machine made by German tractor producer Fendt, Figure 2.22. The machine was actually developed for green forage harvesting. This harvester has no reel, which is to be compensated by the appropriate travel speed. The elevator with a rather high slope is also a weak spot of the machine because the pulling elements cause a high level of friction. One further demerit is that the machine transports material to the processing unit, after filling the hopper, losing time working as a transport vehicle. Another example of a machine, primarily developed for green forage harvest, is a self-propelled mower produced by the company de Pietri, Italy. A smaller type is equipped with an aggressive chains–laths elevator and it is possible to assemble additional cutting knifes. This solution

FIGURE 2.22. Self-propelled mower Fendt "Agrobil" at lemon balm harvest.

seems to be efficient if the material is promptly processed by pressing juices or by distillation or extraction, Figure 2.23.

Fulfilling the requests of MAP producers, the company de Pietri has developed a new concept of the machine with well-designed reel and whole-width band elevator. The producer also offers a high-level tipping type, Figure 2.24. This machine fulfills a lot of listed requests with the usual disadvantage for users in developing countries, that is, with high prices. Even in developed countries the machine is used by groups of farmers, machinery rings, or cooperatives.

Self-propelled machines are always more expensive in comparison with trailed machines. The potential compromise is to use repaired and reconstructed mass-crops agricultural machines, for example, combine harvesters. In that case, the engine, transmission, and chassis should be in good condition, or should be easily (i.e., inexpensively) repaired. Of course, other prerequisites have to be fulfilled, like the possibility of incorporating an adequate conveyor in the proper place. A cutter bar and an inappropriate auger conveyor are the usual elements of a common combine header. Replacement of these parts with a better double knife cutter bar, and band conveyor could be an extensive and expensive reconstruction for some machines. An example of a reconstructed combine harvester is shown in Figure 2.25.

FIGURE 2.23. Self-propelled mower for green forage de Pietri at Echinacea harvest. *Source:* Martinov et al., 2001. Reprinted with permission.

FIGURE 2.24. Self-propelled mower with high-level tipping hopper, De Pietri.

FIGURE 2.25. Example of a combine harvester reconstructed for herbaceous MAP harvesting.

Plescher (1995) reported about one attempt to combine the harvester Claas to the herbaceous MAP harvester. All needed improvements to secure good harvesting quality were incorporated in the machine. This one-man operating machine is aimed at small and medium MAP producers in Germany.

A favorable harvesting machine for small- and medium-scale MAP producers is a trailed machine. This machine has also to fulfill specific demands (Martinov and Tesic, 1995), Figure 2.26.

Here building-in of "torpedoes" is also possible, as well as different types of tines or paddles for reel bars. The use of a hydrostatic drive for cutter bar, reel, and band conveyors is more convenient and nowadays feasible even in developing countries. It means installation of a hydraulic unit powered by PTO with steering possibility. The machine should be positioned to enable coupling of a trailer to the tractor. The tractor's side right position will enable a good view from the driver's position. Changing of machine–tractor position, from travel to work, is needed.

Some old-fashioned mower loaders are still in use. An example is shown in Figure 2.27. This is a typical trailed machine used for green forage. Solid construction enables full function in difficult conditions too. The cutter bar is conventional, the reel does not have the needed adjustment possibilities; the material is transported to the elevator by two side auger conveyors. Harvested material is transported to the trailer by a chain-laths elevator. It is pulled over the metal bottom, which causes considerable friction. Chain-laths elevator leads material to the trailer. All of this has a negative impact on the quality of harvested crop. On the other hand, the machine is simple, the tractor

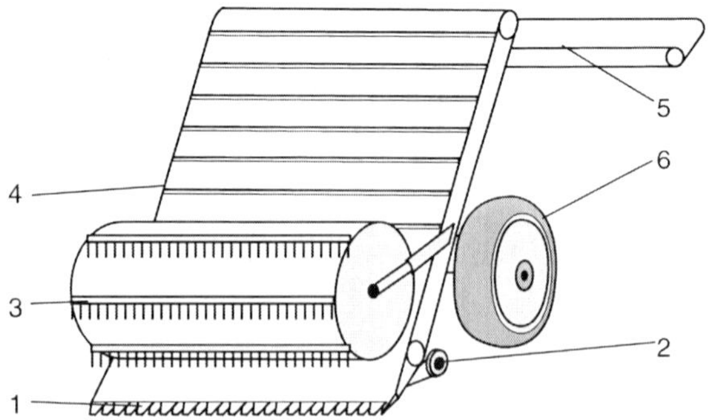

FIGURE 2.26. Concept of a trailed machine for herbaceous MAP harvesting. 1–double knife cutter bar, 2–cutter bar height supporting wheel, 3–reel with height, horizontal position and RPM adjustment, 4–full width band elevator, 5–perpendicular band conveyor, 6–distance adjustable wheels.

FIGURE 2.27. Trailed mower loader by peppermint harvest.

of about 40 kW is sufficient, and it is easy to couple the trailer to the machine. This and similar machines are still compromising solutions for many farmers.

Figure 2.28 shows another machine, also primarily aimed at green forage harvesting and usable for MAP as well. Here there is a double knife cutter bar and a chain-laths elevator with long tines compensating for the lacking reel. The disadvantage is that the elevator has to pull the harvested material over a steel sheet metal slope. After that the material falls into the perpendicular band and it is transported to the bypassing trailer. All machine groups are hydraulically driven with high adjustment possibilities. If the sheet metal slope could be changed to a band elevator the machine would fulfill almost all demands for the successful harvesting of herbaceous MAP.

Figure 2.29 shows a self-loading forage wagon redesigned for the harvesting of herbaceous MAP. The machine is based on the common pick-up self-loading forage wagon with additional knifes for the cutting of stalks. The double knife cutter is driven hydraulically. The reel takes the elevator that brings the harvested material to the load space. The PTO driven floor conveyor enables easy unloading. The disadvantage is that the wagon offers no reloading to trailer. The machine is designed for mass crops, and so carries a price tag that is acceptable on the market. It is a good solution for small and medium farms in developing countries with fields located close to the drying and processing facilities.

FIGURE 2.28. Trailed machine for MAP harvesting.

FIGURE 2.29. Self-loading forage wagon adapted for MAP harvesting, manufactured by Regent.

There are some mower-loader machines, developed by small companies that more or less fulfill listed requirements. The mechanical drive costs less, but complicated transmission is a reason for lower reliability. One example of a successful combined machine is the flower harvester shown in Figure 2.30. This is a machine based on the principle of virtual rotating comb. Here is a slightly adapted picking drum used as a reel. The machine is designed for harvesting of chamomile, but the machine would be much more profitable if used for more crops. In some developing countries, it is quite difficult to

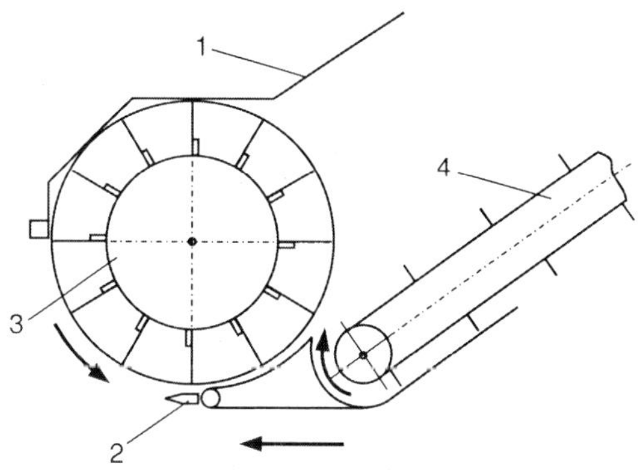

FIGURE 2.30. Herbaceous MAP harvesting machine based on a virtual rotating comb for harvesting of flowers. 1–basic machine for flower harvesting, 2–double knife cutter bar, 3–picking drum acting as reel, 4–band elevator. *Source:* Anonymous, 2003. Reprinted with permission from Euro Prima.

purchase a double knife cutter bar of appropriate width. Usually the width is up to 1.8 m, but to be profitable a mower should have at least 2 m. The nonavailability of hydraulic components, especially slowly rotating hydraulic motors, could also be a problem in some developing countries.

In some countries, MAP could be predried or dried on the field. The success of these procedures depends on many factors: contamination with soil, changing weather, quality requirements, availability of reliable weather forecast, and the like. An example of a windrower used for other crops, with band conveyor mounted perpendicularly, is presented in Figure 2.31. Gentle perpendicular transport of material enables forming of correct swathes. This procedure is used in two-phase harvesting of different MAP.

If the weather conditions are stable, the machine can also be used for peppermint harvest. Of course, the color of the leaves will be changed due to the impact of the weather. This procedure is acceptable if the color change matches the buyers' requests. If production is aimed at producing tea bags, this characteristic is less important. The mowed material will be dried in swathes, on field, without any energy input. After drying it can be collected by using a pick-up header. Threshing and cleaning are very efficient, with high output, but generating a slight reduction of essential oil content and quality. This lower quality is compensated by significant cost reduction. One fur-

FIGURE 2.31. Windrower for swath forming, on field drying of MAP.

ther and serious problem in the application of this procedure is instable weather. Some longer periods of rain at the time of harvest are especially possible during the second and third cuts. Now, good weather forecast services are available everywhere, which makes the realization of this harvesting type more possible.

For some purposes like small plots, testing fields, and the like, a small portable mower can be used. This type of mower, based on the tea leaves pruner and plucker is shown in Figure 2.32. This simple machine can be applied for harvesting at small farms.

For harvesting some special plants additional requirements are to be met. An example is the use of a special cutting apparatus developed for the harvesting of low plants, that is, alpine lady's mantle *Alchemilla alpina* (Stekly et al., 2002), shown in Figure 2.33. The harvester, shown in Figure 2.20, was here used as a basic machine and various mowing devices tested on it. The problem was high content (more than 20 percent) of ash, that is, soil in harvested material. Mowing with hydrostatically driven lawn mowers, and with sickle blades, showed best results. Ash content was reduced to 7.8 percent, which is still very high.

Some attempts have been made to harvest only the leaves, instead of the whole overground plant, in the cases where only the leaves contain active ingredients, for example, peppermint or lemon balm. The device shown in Figure 2.11 and the picker presented in Figure

FIGURE 2.32. Small or test plots portable mower based on the tea plucker–pruner.

FIGURE 2.33. Special mower for the harvesting of alpine lady's mantle.

2.45 were used as strippers (Martinov and Tesic, 1995). The stripping of leaves was successful but a high level of losses of essential oils and negative color changes were recorded. The growing of new plants, for some plants, is not possible due to residual stalks on the field, which is one further problem.

The mechanization of herbaceous MAP harvesting on small plots in arid regions, with stable irrigation infrastructure, as it is shown in Figure 2.34, is a special task and a challenge for engineers and farmers. The solution could be the use of an engine-driven plucker–pruner, as shown in Figure 2.32, in combination with pneumatic transport to the small tractor tagged behind or to the additionally mounted hopper.

Flower Harvesting

The harvesting of flowers is a typical and special problem of MAP production. It is an operation that is seldom needed by other crops. In most cases, the procedure of stripping is used. Stripping has been also used as a harvesting principle for cereals, but, due to certain problems of threshing and straw management, it is not widely accepted in practice.

Innovative attempts in the field of flower harvesting have a long history (Ebert, 1962, 1982). It was a challenge for many generations of innovators, especially for those involved in MAP production. Some developed machines are shown in Figure 2.35.

FIGURE 2.34. Medicinal plant plots with irrigation infrastructure, example from Cyprus.

(a)
(b)
(c)

FIGURE 2.35. Review of design development of the chamomile harvester. *Source:* Reprinted with permission from Müller, 2000.

Picking tools are in almost all cases some kind of comb. The systematization of pickers is presented in Figure 2.36.

Machines with linearly moving combs are used very rarely. One example of a test machine is shown in Figure 2.37. The disadvantage of this type is that it covers a narrow and stationary band of canopy.

This field also abounds in numerous patents for diverse types of harvesting tools and machine concepts. One patent (Figuer, 1977) consists of the linearly moving comb type, which has not been widely used in practice. Two Soviet Union patents (Karacharov and Sabsaj, 1971; Chikov et al., 1980) describe the widely used rotating comb type, with thick-spaced fingers and rotation that is in the counter travel direction. Some of today's commonly used machines are based on these principles. Two other patents (Delaunay, 1975; Tesic et al., 1987) describe the rotating virtual comb types. The second one, with the rotation in travel direction, is widely used in East European countries.

The scheme of a picker-type rotating comb with the rotation in the travel direction is shown in Figure 2.38. The combs (5) are formed so as to have space between the fingers-tines bigger than the stem diameter but significantly smaller than the blossom's head diameter. In some cases, fingers-tines are made of a good quality and precisely manufactured cast. Sections are guided by the swinging section rod (2) copying guide profile (1), so as to have the comb in an approximately horizontal position by crossing flowers horizon band–canopy

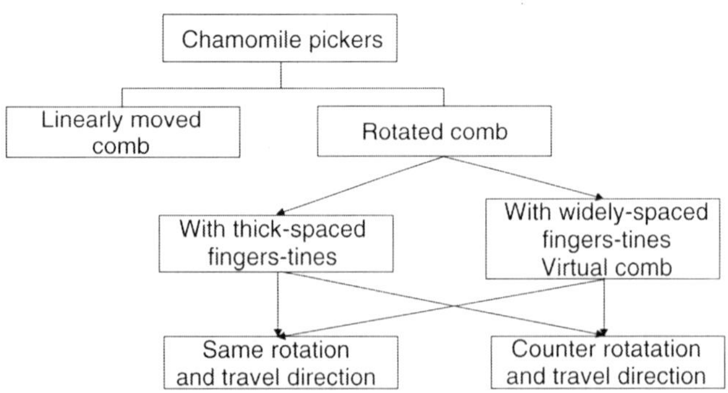

FIGURE 2.36. Chamomile pickers systematization. *Source:* Martinov and Tesic, 1996. Reprinted with permission.

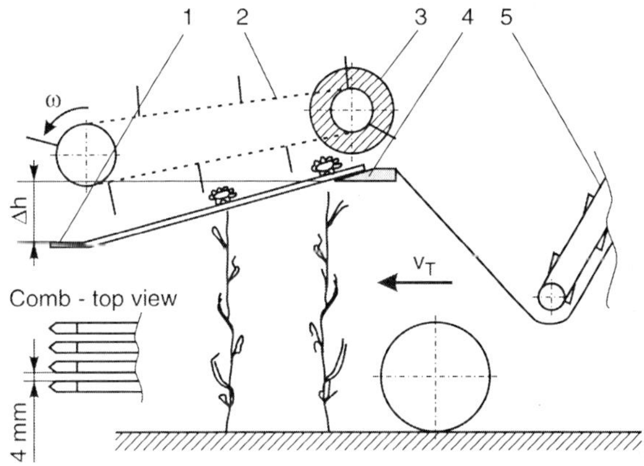

FIGURE 2.37. Chamomile picker with linearly moving comb–prototype. 1–comb, 2–belts with laths and paddles, 3–belts drive, 4–knife, 5–elevator. *Source:* Martinov and Tesic, 1996. Reprinted with permission.

and to be in envelope tangent position before entering position of stems cutter (7). After the cutting of the stems the brushes (6) are activated–rotated, driven by friction between brushes rolls (3) and driving support bar (4). Flowers are scraped from combs and fall down, through hopper (9) to the central positioned band conveyor (10), and are laterally transported to the bin.

Another rotating comb flowers picker is shown in Figure 2.39. This is of the same type with different designed details of mechanism and comb.

The positioning of combs (3) using rolls (2) guided via profile (1) is also applied here. The combs design principle is the same as for type I. Flowers fall down under brush (5), and are axially transported by auger (8) to the side band elevator (9).

Essentially the same type of picker, but with comb, cutter, and brush designed differently, is schematically presented in Figure 2.40. The design is simplified, so that the section of combs (2) is made of steel tins, and fixed to the rotating drum (1). Axially situated counter rotating cutter (3) with spiral knifes is at tangents to the comb's orbit. A rotating brush that scrapes the flowers is placed in the comb. The

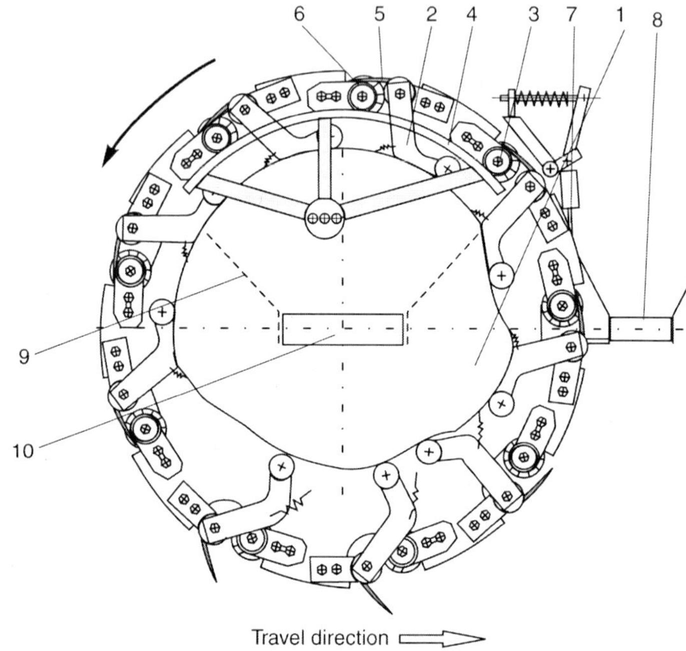

FIGURE 2.38. Rotating flowers picker comb, rotation in travel direction, type I. 1–guide profile, 2–comb section rod, 3–brushes rolls, 4–rolls driving support, 5–comb section, 6–rotated scraping brushes, 7–stems cutter–knife, 8–stems band conveyor, 9–hopper, 10–flowers band conveyor. *Source:* Hecht et al., 1992a. Reprinted with permission.

brushes direct the flower's toward the auger conveyor (5). This solution is easier for manufacturing in small workshops, too, in developing countries.

Due to efficient cutting of stems, high quality of harvested material can be achieved with all three picker types, that is, flowers with no or short stems, but if the plants have intensive branching this advantage is significantly reduced. In that case the location of the branching has the same effect as the flowers, and the longer stems remain. The picking is complicated from an engineering point of view and needs advanced manufacturing technique, not available in all developing countries.

Figure 2.41 shows a self-propelled chamomile harvester that is using picker type I. Here is detail of a picker shown during harvesting.

The details of a rotating comb picker, type II (Figure 2.39), are shown in Figure 2.42. A self-propelled machine using this type of picker, in Figure 2.43, is shown involved in chamomile harvesting.

The rotating comb picker type III (Figure 2.40) and a self-propelled harvester that uses this picker are shown in Figure 2.44. This machine, developed and produced in Slovakia, presents a good solution for medium and big producers of chamomile. The demerits are use of an auger conveyor that causes squeezing of harvested material and manual unloading of the hoppers in the back of the machine that are filled pneumatically.

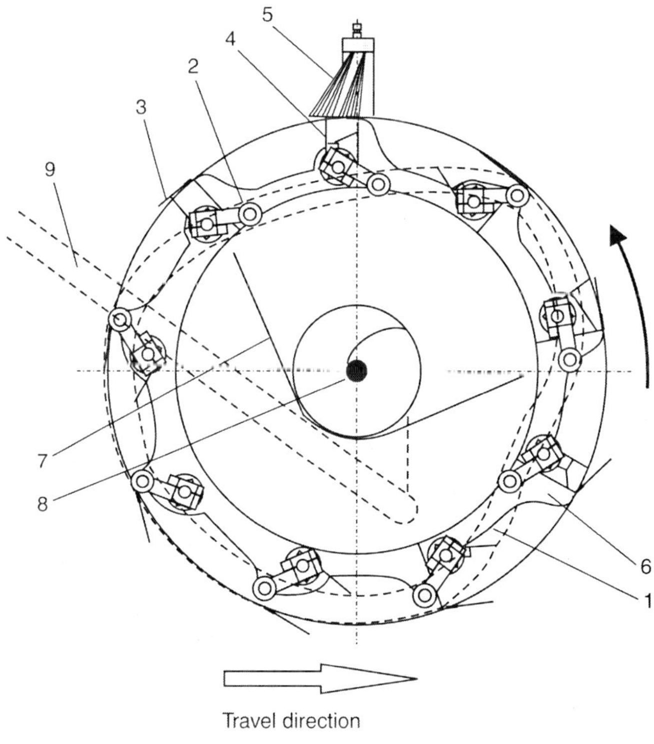

FIGURE 2.39. Rotating flower picker comb, rotation in travel direction, type II. 1–guiding profile, 2–roll, 3–comb, 4–mold plate, 5–brush, 6–cross beam, 7–tub, 8–axial auger, 9–side band elevator. *Source:* Hecht et al., 1992a. Reprinted with permission.

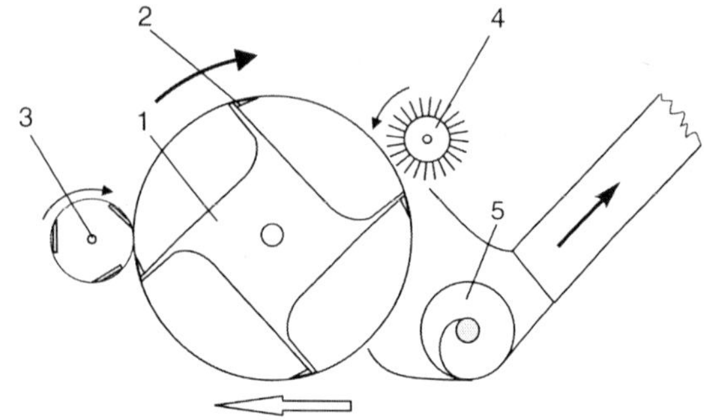

FIGURE 2.40. Rotating flower picker comb, rotation in travel direction, type III. 1–rotating drum, 2–comb sections, 3–rotating cutter with spiral knifes, 4–scraping brush, 5–auger conveyor.

The next flower picker is a virtual rotating comb type, Figure 2.45. It consists of 12 or 16 laths with tines situated at distances of 40 to 60 mm (2) connected to the central cylinder (1). The diameter of the picker rotor is 600 mm or 720 mm. The tines of laths are displaced for a certain longitudinal distance, so that rotating the tines generates a combing effect. The rotation is in the same direction as the travel direction, and a counter type has only seldom been applied. Almost the same type of picker was developed for green bean harvesting. The tests of this comb type for chamomile were successful and the harvester was developed further. Very useful and effective improvement was done in the 1980s (Martinov et al., 1992) and patented (Tesic et al., 1987).

The plants are caught by the tines and pulled into the picking space, between cylinder (1) and bottom sheet plate (4). There occurs intensive combing due to a virtual comb effect with frequent impacts of laths with tines (2) with a special positioning pattern. The picked flowers are thrown toward the band elevator (5). It is important to position the elevator so that the paddle velocity is not directed toward the picker drum, as shown in Figure 2.45, because this causes backflow of harvested flowers and losses. That is why the end of the bottom sheet plate has to be extended to the point T'. This reduces the paddle radius to the needed minimum and eliminates ventilation effects and loss of harvested material.

(a)

(b)

FIGURE 2.41. The rotating comb picker in chamomile harvesting Hege. (a) self-propelled machine, (b) detail of picking drum. *Source:* Reprinted with permission from Müller, 2000.

The combination of rotation of the picking drum (velocity of tines top v_O) and travel speed v_T results in a cycloid trajectory which is, for chamomile harvesting parameters, $v_T = 0.4\ ms^{-1}$ and $\omega = 13\ s^{-1}$, shown in Figure 2.46. Here is also presented a cultivated chamomile plant with the height of about 0.5 m. O-G are the positions of the picking drum center, O'-G' positions of the tine top, and O"-G" of the bending tube. Here two plants, I and II, are presented. The plant is first pulled into the working space and combed by spirally mounted

FIGURE 2.42. Version of a rotating comb picker type II schematically shown in Figure 2.39; details of comb sections.

FIGURE 2.43. A self-propelled chamomile harvester with a rotating comb picker type II, Linz. *Source:* Reprinted with permission from Müller, 2000.

tines during more than 4.5 drum rotation, before it is again pulled out. The frequent passing of the tines through the material layer generates a combing effect and causes the stripping of the flowers.

The stalk's outgoing phase, position I", with the angle β to the horizon causes opposite plant stalk mowing, leaving working space.

(a)

(b)

FIGURE 2.44. Self-propelled chamomile harvester with a rotating comp picker type. (a) machine view, (b) detail of drum, scraping brush, and auger conveyor.

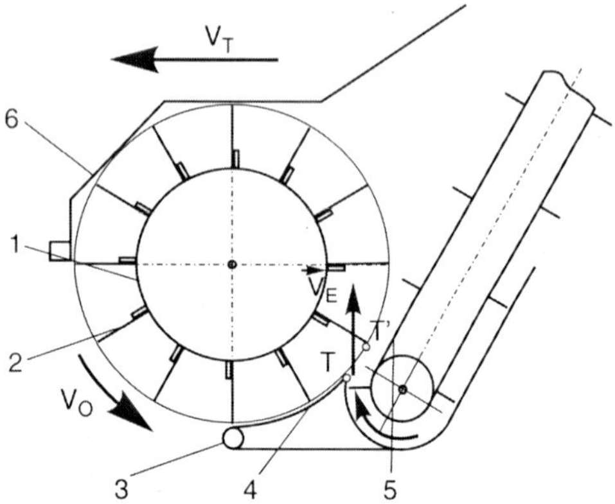

FIGURE 2.45. Virtual rotating comb flowers picker. 1–central cylinder, 2–laths with tines, 3–bending tube, 4–bottom sheet plate, 5–band elevator, 6–housing.

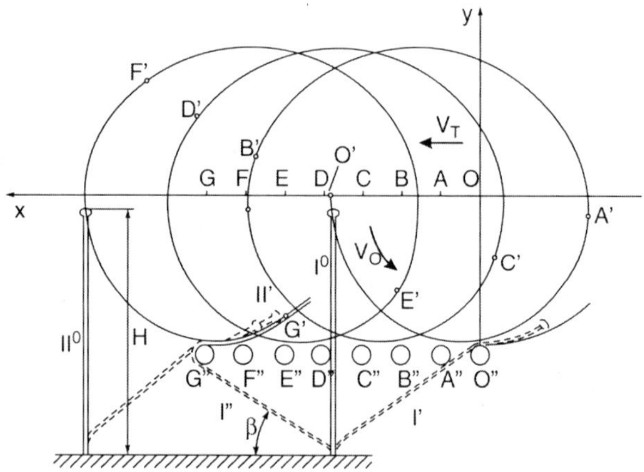

FIGURE 2.46. Flowers stripped by a virtual rotating comb picker type.

It is assumed that it causes the throwing out of some short stalk flowers generating losses. It can be reduced by the slowing down of the stalk velocity in this phase, that is, travel speed. That is why the maximal travel speed of the harvester is limited to 0.5 ms^{-1}.

Figure 2.47 shows the side-trailed machine with a virtual rotating comb picker type in chamomile harvest. There are a lot of possibili-

FIGURE 2.47. Chamomile harvesting using a rotating virtual comb picker type, BK-83.

ties to incorporate a virtual rotating comb picker type in a simple machine design as in BK-83. The scheme of the mounted harvester is presented in Figure 2.48. Depending on the working width, 1.2 or 1.5 m, the machine could be used with a tractor of 30 or 40 kW, respectively. That is more a question of hydraulic hitching capacity and tractor stability than PTO power; for one meter of working width about 5 kW of PTO power is needed. For trailed machines the weight of the harvested material has to be included in the calculation of the tractor stability. The biggest influence on this calculation is the expected yield and field length, that is, positions of unloading. This type of harvester is a possible solution for small and medium producers, especially for smaller plots and hilly regions. One worker is needed to follow the machine, to control harvesting, and to reload the harvested material from the rear to the side basket. Reloading to the trailer is also manual. Figure 2.49 shows a tractor mounted with a 1.5 m working width harvester in pyrethrum harvesting in Chile.

In all cases of flower harvesting the results of the harvest are flowers with different stem lengths. Further processing, as well as material classification, is influenced by the length of the stem. It is advantageous to separate plants according to the length of the stem, if possible, during harvesting (Martinov and Oluski, 1998). On-machine, that is, on-harvester separation is more efficient than an

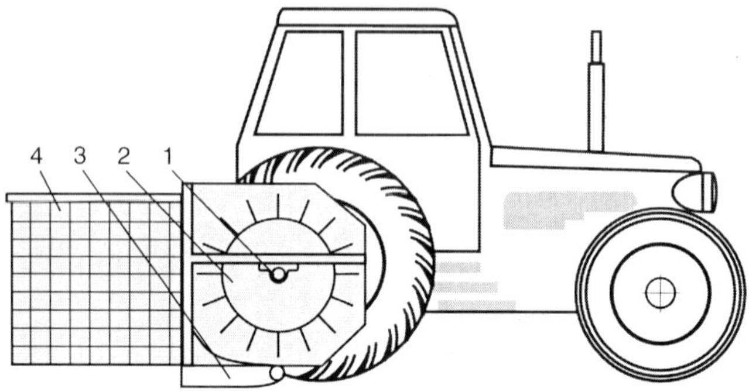

FIGURE 2.48. Tractor mounted flower harvester with a virtual rotating comb picker type. 1–tractor PTO driven shaft, 2–picker, 3–slider, 4–basket.

FIGURE 2.49. Tractor mounted with 1.5 m working width flowers harvester in the harvesting of pyrethrum in Chile.

additional one, due to the continuous flow of plants, whereby these are not interwoven. There are some designs that enable this procedure. Figure 2.50 shows a self-propelled harvester with a rotating comb picker type and integrated sieves for initial separation of harvested material according to the length of the stems. The sieve length is limited due to design reasons; that is why a worker has to control the efficiency of classification, pulling the material backward, and preventing obstruction to the sieve.

(a)

(b)

FIGURE 2.50. Self-propelled flowers harvester with built-in sieves for separation according to the length of the stems. (a) machine at harvest, (b) built in siever. *Source:* Reprinted with permission from Müller, 2000.

Another solution, used for a trailed harvester with a virtual rotating comb picker is schematically shown in Figure 2.51. Here the chassis and body of the agricultural trailer are used as a base for assembling the picking unit. Half of the trailer body is covered with the sieve, driven by tractor PTO, over belt transmission.

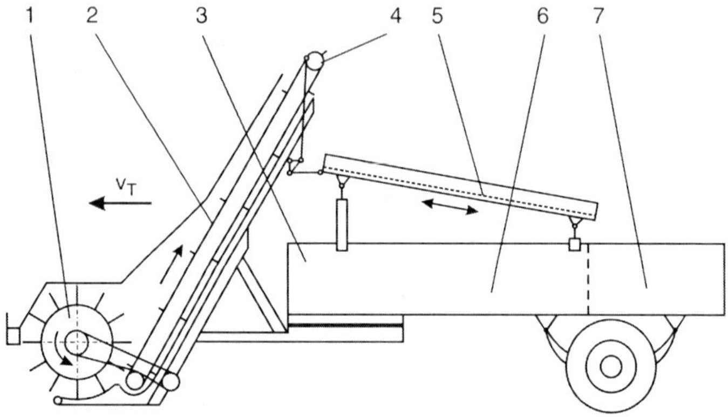

FIGURE 2.51. Trailed flowers harvester with built-in sieve. 1–picker, 2–band elevator, 3–sieve, 4–trailer body, 5–sieve drive transmission. *Source:* Reprinted from Martinov and Müller, 1990.

On-machine separation of flowers with different stem lengths contributes to higher quality of material and enables better drying as well. First and second-class material, Figure 2.52(a), are to be dried promptly and carefully. This is on average 60 to 80 percent of the total harvested material. Other material, with longer stems, can be stored for a few days in a ventilated room, and dried after the completion of drying of the first and second class material.

Figure 2.52 illustrates different chamomile: (a) on-machine separated and (b) unseparated material.

The amount of flowers with short stems is approximately the same for both materials shown in Figure 2.52, but in the case of nonseparated material short stemmed flowers are dropped down through layers of long stemmed material and are not visible, which gives an impression of a bad harvest.

The success of the harvest depends on numerous influencing factors. The crop properties are the most influencing. Some of them follow: the height of the plants, flowers horizon band, branching, mechanical characteristics of stems, plant height uniformity, weed population and uniformity of maturation. That is why a comparison of harvest results can only be an orientation for the evaluation of machines. Different types of pickers and harvesters are difficult to com-

(a)

(b)

FIGURE 2.52. Examples of harvested chamomile flowers. (a) separated (Hege, Figure 2.41), (b) not separated (BK-83, Figure 2.47). *Source:* Reprinted with permission from Müller, 2000.

pare, especially under different harvesting conditions. The list of significant harvesting outputs for three machines is given in Table 2.1. All three machines have been tested in local, noncomparable crop conditions, and the data can be used only to get an impression about some characteristics. Hege and BK-83 have a 2 m working width and Linz has 3 m. The difference in quality of the harvested material, between Hege and material harvested by Linz and BK-83, is the result of the built-in sieve in Hege. Much lower losses with Hege can be a result of the narrow horizon of flowers. Actually, total losses of up to 15 percent are acceptable, especially if there is a wider horizon of flowers and higher weed population in the crop. The quality comparison of the Hege and Linz harvesters was done by Hecht et al. (1992a), and results similar to those shown in Table 2.1 were obtained. They also made the comparison with manually harvested material. The material harvested manually had lower losses, but the length of the stems was longer on average.

Evaluation of the quality of the material depends on standards and market requirements, which are even more important. Generally, chamomile harvesting is evaluated positively if the results are more than 60 percent of flowers with stems shorter than 4 cm (about 2 cm after drying). It can be concluded that the mechanized harvesting of chamomile is basically solved. It can be improved, but it is not an acute engineering task any more. For other plants, for example, marigold, yarrow, and pyrethrum, additional efforts, combinations of engineering and breeding efforts, are needed. Selective maturation, difference in plant height, mechanical characteristics of stems, destruction of blossoms, and others are to be thoroughly investigated. Hecht et al. (1992a) reported some results in marigold and yarrow harvesting, but the experiment was performed for small plots and the data cannot be used for practically usable conclusions. Martinov and Adamovic's (2002) preliminary results of mechanized harvesting of pyrethrum using the machine shown in Figure 2.49 gave promising results. The amount of flowers with stems shorter than 2 cm was over 45 percent, shorter than 4 cm over 60 percent, and total losses were 9

TABLE 2.1. Significant characteristics of three flowers harvesters–chamomile harvest (Müller, 2000).

	Harvesting capacity, ha/h	Labor needed, h/ha	Flower, %	Herbs, %	Losses, %
Linz, Fig. 2.43	0.4	2.5	72	13	15
Hege[1], Fig. 2.41	0.25	8.0	90	7	3
BK-83, Fig. 2.47	0.3	3.3	73	14	13

Source: Reprinted with permission from Müller, 2000.
[1]Machine with built-in sieve, that requires higher labor needs and lower capacity, but resulted in higher quality of harvested material.

percent. The need to cut the stems before extraction is identified as a problem. Mechanized pyrethrum harvesting is practiced in some countries (Tasmania), but other procedures, i.e., mowing of mature flower canopy, are also performed. It needs additional processing of harvested material. This procedure has not been published and it seems to be maintained as a producer's secret.

Fruit and Grain Harvesting

Fruit harvesting is usually not mechanized. Many plants, such as dog rose *(Rosa canina)*, apple rose *(Rosa villosa)*, blackthorn *(Prunus spinosa)*, hawthorn, and others are commonly harvested manually. The efficiency is usually very low, for example, 5 kg/h per picker, and the quality is inhomogeneous. Some attempts to cultivate wild plants resulted in higher yield, quality and harvesting efficiency. A typical example is the cultivation of the sea buckthorn *(Hippophae rhamnoides)*.

The use of berry fruits, i.e., currants, harvesters with rotating tines, for the harvesting of row-cultivated dog and apple rose, offers the possibility to mechanize this process. But, due to the fruits' high holding forces, the losses are usually very high, up to 50 percent (Triquart, 1997a,b). The holding forces for some fruits are given in Table 2.2.

The cutting and removal of fruit branches makes picking of fruits easier. This two-phase procedure still needs a lot of labor, but is certainly an improvement in harvesting efficiency. This is more efficient in the case of field production and less for spontaneous flora. The re-

TABLE 2.2. Holding forces for some fruits, fruit-branch/stem connection.

Fruit	Mass (g)	Holding force (N)	Theoretical acceleration needed for removal (ms^{-2})
Sea buckthorn (*Hippophae rhamnoides*)	0.3	1.4	4700
Apple rose (*Rosa villosa*)	5.4	8.9	1600
Common elder (*Sambucus nigra*)	0.15	0.3	1700
Blackthorn (*Prunus spinosa*)	2.0	4.3	2100

Source: Triquart, 1997a,b. Reprinted with permission.

moval of fruits from branches can be done by manual picking, including use of any hand tool to make it easier, or mechanical threshing. A tractor trailed machine for the threshing of the branches is schematically presented in Figure 2.53. The machine is pulled between the rows of plants and the cut branches are deposed on to the conveyor (7). The tool (8), shaker, removes the fruits. The other parts, the specially designed conveyors and elevators, are aimed to remove the impurities. All parts are driven hydrostatically via tractor PTO.

The shaker is shown in Figure 2.54. It consists of a shake drive (3) with tines (1). The branches are positioned at R_O and shaken by oscillating tines. Oscillation frequency, i.e., the intensity of the shaking, can be adjusted.

In most of the cases the frequency of 25 to 30 Hz, with an amplitude of 16 mm, was used for the removal of fruits. The travel speed of the machine was 250 to 310 m/h and the capacity 0.21 to 0.26 ha/h. The low capacity was compensated by good threshing efficiency and low losses. The losses are different for diverse plants but much lower than for the harvest with the berry fruits harvester. For example, the losses were 1.5 percent for the harvesting of apple rose using this machine and 20 percent with the berry fruits harvester (Triquart, 1997a,b). Higher labor costs are compensated by lower losses, even in developed countries. The same machine used for threshing was tested in the harvesting of the sea buckthorn (Gätke and Triquart, 1992). The results were evaluated as positive: the capacity was 0.08

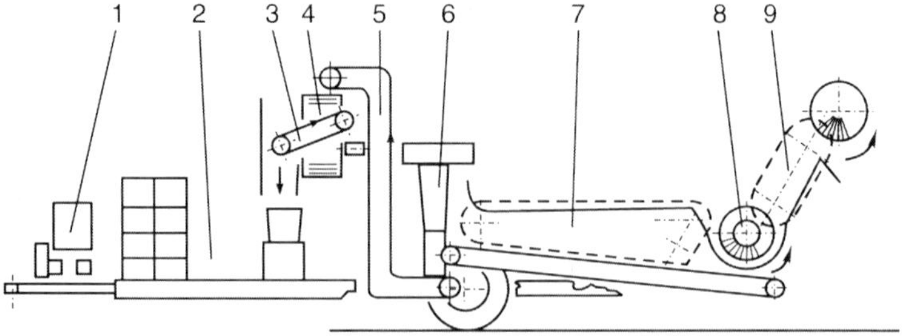

FIGURE 2.53. Machine for threshing and cleaning of berry fruits, blackthorn. 1–hydrostatic drive, 2–workers platform, 3–fine part separation band elevator, 4–sieve chain, 5–cup conveyor, 6–leaves separator, 7–receiving conveyor, 8–shaker, 9–branch ejector. *Source:* Triquart, 1997a,b. Reprinted with permission.

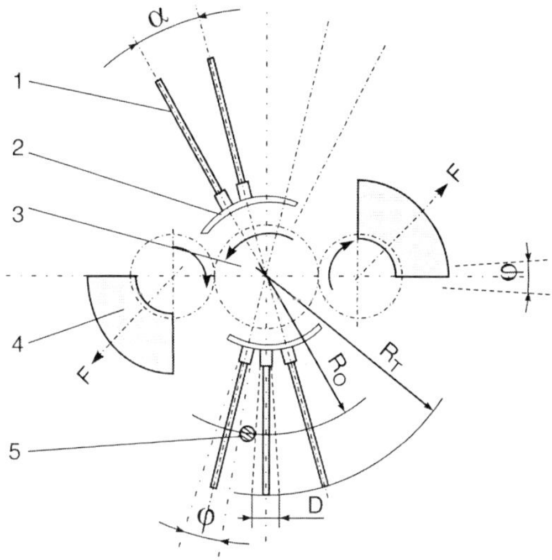

FIGURE 2.54. Shaker. 1–tines, 2–tines' drum, 3–shake drive, 4–unbalance mass, 5–plant branch, α–tines distribution angle 15°, φ–oscillation angle 8,6°, D–shake distance, R_T–tines radius (300 mm), R_O–operating radius, F–centrifugal force. *Source:* Triquart, 1997a,b. Reprinted with permission.

to 0.1 ha/hour and the losses approximately 10 percent. The use of the machine described above or similar threshing machines seems to be a promising solution for the mechanized harvesting of fruits.

For harvesting of grainy MAP slightly adjusted common cereals combine harvesters are usually used. There are yet some problems and there is a need for special adjustments to be made. Mechanical, morphological, and other plant properties, as well as plant architecture, have an influence on the efficiency of the combine harvester units: header, thresher, cleaner, straw walker, and the rest. The losses generated in the combined separation unit are important for evaluating the harvest. The following plant and grain characteristics have the biggest impact on the harvesting procedure:

- Corn ear binding force
- Grain and chaff size and shape
- Grain and chaff airflow characteristics
- Uniformity of maturation
- Moisture content of grains at harvesting time
- Length and mechanical characteristics of straw

- Straw moisture content at harvesting time
- Corn horizon in crop

The "big" companies are not interested in R&D of harvesting procedures and the technique of MAP. They tested their combines in MAP harvesting in cooperation with public institutions and as a kind of sophisticated advertising. Typical examples of combine harvester parameters for MAP harvesting, as a result of such effort, could be obtained from Anonymous (1988). Using these data farmers should be able to adjust other types of combine harvesters.

There are many other influences on the harvesting procedure. Hecht et al. (1992b) investigated the use of the combine harvester in some grainy MAP harvesting using diverse types of combine harvesters: Deutz-Fahr, Massey Ferguson, Claas, and Fortschritt. Hecht et al. (1992) concluded the following:

- No major reconstructions of the combine harvester are needed. For some plants the header table has to be extended and the divider cutter bars used, as it is usually applied for rape harvesting.
- The combine harvester adjustment is strongly dependent on the plant type and conditions. The data are different for different types of thresher and cleaning units of the combine harvester.
- Fennel is characterized by a late harvesting period, high straw moisture content at that time, 20 to 60 percent, and high and strong stalks. The adjustment is similar to wheat and the divider cutter bars make the harvest easier.
- It is easy to thresh coriander, but because of the weak grain–earn bond, the harvest has to be realized before full maturity. At that time the straw is also moist, which can cause drum wrapping.
- Caraway is easy to thresh. The losses are lower due to strong grain–earn bonds. The combine harvester parameters are similar to parameters for other cereals. The problem is unequal ripening.
- Milk thistle is characterized by a high level of grain failure losses. It is very important to find a compromise between these losses and the harvest of grains that are not yet ripe. Moist

grains with hairs cause problems for the threshing apparatus and the cleaning device, but on the other hand, late harvest causes up to 40 percent failure losses. The decision making is harder due to the unequal ripening of seeds. It is useful to divide the cutter bars. The reel tines should be unaggressive or soft paddles should be used.

- In poppy seed harvesting the adjustments are done in accordance with very small size of the seed. The problem is that the crushing of the capsules provokes seed crushing, i.e., losses. Hence, the feeding conveyors should be as unaggressive as possible.

Losses are the most significant characteristics of harvesting success. Losses measured by MAP harvesting using combine harvester are shown in Table 2.3.

There are other test results that provide data usable for MAP producers (Alpmann, 1990). An example of results of the combine harvester Fortschritt E 514 are presented by Setitz and Paulus, 1988. The data given here can be used only as orientation.

Some very usable guidelines for milk thistle harvesting are given by Oheimb (1990), where the reduction of harvest losses has been focused on. The appropriate harvest time for one phase of the harvest has been described, "when the upper leaves are getting brown". Otherwise wet leaves, i.e., top hairs, cause obstruction to the sieves. The following general instructions have been given:

TABLE 2.3. Combine harvester losses by harvesting of MAP.

Plant	Losses, %	Location of major losses	Comment
Fennel	2.2-60.0	–	High straw and grain moisture
Coriander	1.5-2.9	–	Straw coiling
Caraway	6.5	–	Good harvest of mature grains
Milk thistle	34.0	Thresher, 1/3 cutter bar	Sieve and thresher blockage
Poppy seed	1.9-4.0	Header, auger	Destroying by conveyors
Mustard	5.1-12.0	–	–

Source: Data compiled from Hecht et al., 1992b.

- Cutter bar–standard, used for cereals
- Cutting height–one-third over ground
- Reel–teens covering used for sunflower
- Drum RPM–approx. 800
- Concave and sieves–as for cereals
- Blower–three fourths of full capacity

Moisture content of grains at harvest time was about 30 percent, and combined travel speed from 1.5 to 4.0 km/h, depending on crop conditions.

Another grain harvest procedure is mowing and windrowing with the machine shown in Figure 2.31, or similar, as a first phase. The second phase is the pick up of dried material by using a combine harvester equipped with adequate header. This procedure generates lower losses, but additional machine and combine harvester header are needed.

Harvesting time influences the yield of grains and essential oils. The data for fennel harvesting in phases one and two are given in Figure 2.55. Here, the harvest time is much lower if a two-phase harvest was applied.

As is the case for cereals and other crops, harvesting of lodged (laid down) crop causes problems even in the case of MAP. It would be manageable if all plants were to lie in one direction for the harvest can then be successfully realized by mowing in the counter direction.

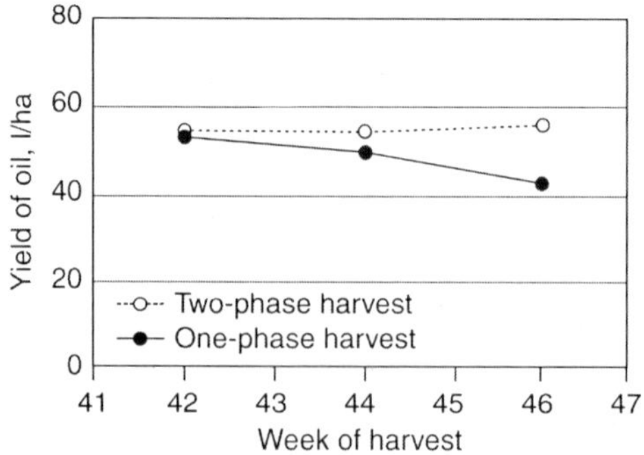

FIGURE 2.55. Essential oil yield of fennel after one-phase and two-phase harvest vs. harvest time. *Source:* Müller, 2003. Reprinted with permission.

In the opposite case, the harvest would incur much higher losses. One example of such a harvest condition is shown in Figure 2.56.

Mechanized harvesting of grainy MAP could be considered as solved. Some developments and further improvements are possible, especially in the direction of reduction of losses. This is a typical multidisciplinary task, for engineers, breeders, and agronomists.

Harvesting of fruits, for example, dog roses, is still not mechanized. Some improvements in manual wild crafting could be achieved by using pickers or similar devices. Actually, it is more likely that some new varieties for cultivated production, with adequate plantarchitecture and other characteristics suitable for mechanized or semi-mechanized harvest, will be developed.

Special Harvesting

There are some procedures and machines for certain plants that could be included in MAP that were developed in the past two decades. Two typical examples are presented here.

Pumpkins, especially the varieties with high oil content, have recently become interesting products for macrobiotics, food, oil production, and as raw material for the pharmaceutical industry. Some harvesting procedures are possible for this plant. Due to the negligi-

FIGURE 2.56. Harvest of mustard lying in different directions. *Source:* Müller, 2003. Reprinted with permission.

ble nutrition level of the plant meat, harvesting of seeds is dominantly practiced. The transport of the whole pumpkin from the field to the farmyard is not economically justified due to the high mass. There are, though, some rare exceptional cases when the meat can be efficiently used as forage. That is why, in most of the cases, the pumpkins are destroyed on the field after the seeds separation. There are some machines developed for this operation. A typical one is shown in Figure 2.57, where there is a large drum mounted sideways, with fingers that pick up and elevate the pumpkins. A bar structure takes off the pumpkins from the fingers and guides them to the crushing drum. The body of the pumpkin is destroyed after which the seeds are separated. It is necessary that the pumpkins are put in a band not wider than the pick-up device before the machine starts picking up. This is realized either by a tractor-mounted scraper or manually. Some harvesting machines operate without a pick-up device, so the pumpkins have to be manually fed to the machine hopper. The procedure afterward is the same.

Some plants that were earlier used for various purposes, for example, flax or hemp used for fibers, are now treated as medicinal and aromatic plants. Flax oil is used in the pharmaceutical and cosmetics

FIGURE 2.57. Special machine for pumpkin seeds harvesting.

industry. In the past, the flax harvesting machinery was oriented toward successful harvesting of stalks, that is, raw material for fibers. In that case, or in the case of seed material production, the grains have to be removed. An example of a flax harvesting machine with a built-in thresher for grains is shown in Figure 2.58. The thresher consists of a drum with tines and operates by stripping grains from earstalks.

It is expected that many more plants have medicinal properties that have so far not been identified. With the introduction of new products, that is, new MAP, a new market is expected to be developed. Entering this new market should be, at least in the beginning, a lucrative activity. Of course, mechanization of harvesting could have a crucial role for profitability. This will be a great challenge for inventive farmers and engineers.

TRANSPORT

The transport of harvested material has a huge impact on material changes and on the quality of the final product. The best preservation of the quality of harvested plants would be prompt drying or other

FIGURE 2.58. Flax harvesting machine with built-in thresher for grains.

kinds of processing. That is only possible, for example, with on-field drying.

The duration of transport should be reduced to less than three hours. This is only a general recommendation, because the effects of quality degradation depend strongly on the material's moisture content and air temperature. If the temperature and moisture content are higher, the transport duration has to be shorter. The negative effects of postponed drying depend on the material's properties, batch density, layer height, and the like. High moisture content and temperature are good prerequisites for intensive aerobic and anaerobic bacterial activities. Bacterial activity generates warming-up of plant material that can be measured. The increase of temperature can be very high, even up to self-ignition. All processes of moisture–heat effects are not entirely explained, but the negative impact to the plant material quality has been proven.

The results of a test of microbial count change during transport are shown in Figure 2.59. This testing has been done for a transport duration of an average of 78 min at temperatures from 24 to 35°C. It was concluded that the rise in temperature of only 3 to 4°C was negligible, due to the short duration of the transport. An average rise of microbial count from 4.4×10^4 cfu/g to 2.3×10^5 cfu/g was recorded. It was also detected that the number of bacteria types in samples before the transport was four and after the transport six. In almost all samples, after transport, the Bacillus type was detected. Only the content of the molds has been reduced during transport.

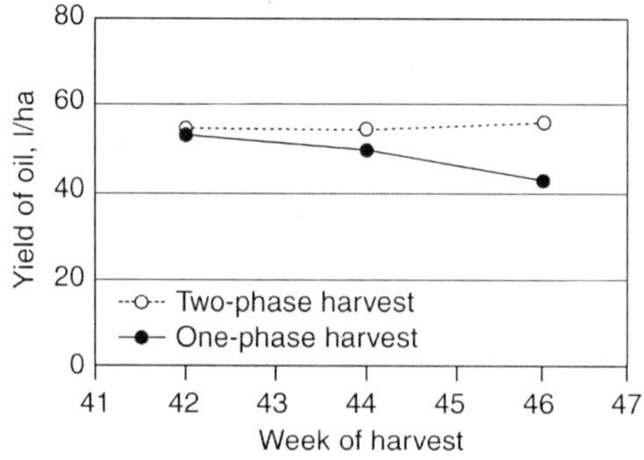

FIGURE 2.59. Changing of microbial count during transport-example St. John's wort. *Source:* Graf et al., 2001. Reprinted with permission.

In some cases, a longer transport is required for to various reasons. To overcome the problem of microbial count increase harvest and transport could be achieved during the colder part of the day or by night. Ventilation of transported material by using special trailers can also help preserve quality, but this causes additional costs for producers.

REFERENCES

Alpmann L. 1990. Ernte von Grassamen, Ölrettich, Rübsen und Senf, In Mähdruschernte von Sonderfrüchte. KTBL-Arbeitspapier 139. Kuratorium für Technik und Bauwesen in der Landwirtschaft e.V. (KTBL): Darmstadt; 97-101.

[Anonymous]. 2003. Production palet. Available from www.europrima.co.yu.

Bomme U. 1976. Kartoffel-Vollernter für Baldrian. Landwirt, (8), 26 pp.

Bomme U, Eberlein H, Rödel G. 2000. Grünguternter für Heil- und Gewürzpflanzen optimiert. Schule und Beratung, 07(IV): 10-15.

Chikov PS, Pakaln DA, Martynov, JF, Perebejnos, VS, Jarmashev, JN, Ljah, AA, Dushenko, AP, Borodin AI. 1980. Mashina dlja uborki socvetnij lekarstvennyh rastenij. USSR Patent. No: 759069. 2 pp.

Dachler M, Pelzmann H. 1999. Arznei und Gewürzpflanzen. Oestereischen Agrarverlag, Klosterneuburg: 353 pp.

Delaunay HMJ. 1975. La machine à cueillir la camomolle. France Patent. No: 2297554. 4 pp.

Ebert K, Schubert H. 1962. Kamille-Ernte voll mechanisiert. Deutsche Apotheker-Zeitung; 102(6): 167-168.

Ebert K.1982. Kamille, Echte, in Arznei- und Gewürzpflanzen; Ein Leitfaden für Anbau und Sammlung. Wissenschaftliche Verlagsgesellschaft, Stuttgart.

Figuer GRL. 1977. Maschine für Kamillen Blütenernte. West German patent (priority Spain). Patent no: 2546539, 11 pp.

Gätke R, Triquart E. 1992. Maschine zur mechanisierten Ernte von Sanddorn im Schnittverfahren, Gartenbaumagazin; 1(9): 57-58.

Graf C, Martinov M, Müller J. 2001. Impact of transport and drying processes on microbial count of medicinal plants: Example St. John's wort (*Hypericum perforaturm* L.). Medicinal Plant Report; 8(8): 7-16.

Hecht H, Mohr T, Lembrecht S. 1992a. Mechanisierung der Blütendrogenernte–Vergleich zweier Pflücksysteme. Landtechnik; 47(6): 276-280.

Hecht H, Mohr T, Lembrecht Silvia. 1992b. Mähdruschernte von Körnerdrogen. Landtechnik; 47(10): 494-496 and 504.

Karacharov GM, Sabsaj VD. 1971. Mashina dlja uborki socvetij lekarstvennyh rastenij. USSR patent. No: 322148. 2 p.

Martinov M, Müller J. 1990. Produktion von Kamille in Jugoslawien. HGK Mitteilungen; 33(5): 45-47.

Martinov M, Tesic M, Müller J. 1992. Erntemaschine für Kamille. Landtechnik; 47(10): 505-507.

Martinov M, Tesic M. 1995. Mechanization of harvest, drying and primary processing of medicinal plants and plant products. Lekovite sirovine; 44(14): 43-55.

Martinov M, Tesic M. 1996. Mechanization of chamomile harvesting and processing. Medicinal Plant Report; 3(3): 38-51.

Martinov M, Oluski L. 1998. Machines and equipment for chamomile production–Twenty years after. Medicinal Plant Report; 5(5): 37-49.

Martinov, M., Müller, J. Raicevic, D., Erzegovic, Dj.1999. Machine and equipment for harvesting, drying and primary processing of sage. In D. Brkic Zalfija (*Salvia officinalis* L.). Institut za proucavanje lekovitog bilja "Dr Josif Pancic" and Art Grafik, Belgrade.

Martinov M, Müller J, Veselinov B, Konstantinovic M. 2001. Harvesting of foliage medicinal and aromatic plants: Review of Machines and Procedures. Medicinal Plant Report; 8(8): 23-33.

Martinov M, Adamovic D. 2002. Mechanised harvesting of pyrethrum (*Pyrethrum cinerariaefolium* Trev.): First Testing. Medicinal Plant Report; 9(9): 23-27.

Müller J. 2000. Nachentetechnologie–Arzneipflanzen Teaching material University of Hohenheim. Institute for Agricultural Engineering in Tropics and Subtropics. Stuttgartt.

Müller J. 2003. Stand und Forschungsbedarf bei der Erntetechnik von Arznei- und Gewürzpflanzen, Arznei und Gewürzpflanzen; 8(2): 56-60.

Oheimb von R. 1990. Ernte von Marientdisteln. In Mähdruschernte von Sonderfrüchte, Kuratorium für Technik und Bauwesen in der Landwirtschaft e.V. (KTBL), Darmstadt: 105-111.

Plescher, A. 1995. Entwicklung und Prüfung einer selbstfahrenden Erntemaschine für Arznei-, Gewürz-, Aroma- und Färberpflanzen. Pharmaplant, Artern: 24 pp.

Sauter GJ, Kirchmeier H, Geischeder R, Rödel G. 2002. Roden von Meerrettich. Landtechnik; 57(4): 204-205.

Schurig M, Rödel G. 1990. Ernte von Arznei- und Gewürzpflanzen. In Mähdruschernte von Sonderfrüchte, Kuratorium für Technik und Bauwesen in der Landwirtschaft e.V. (KTBL), Darmstadt: 101-104.

Setitz P, Paulus S. 1988. Optimierung von Mähdruschverfahren zur Körnerdrogen, Hessisches Landesamt für Ernährung, Ladwirtschaft und Landentwicklung, Kassel.

Stekly G, Rödel G, Eberlein H, Dallinger J, Schneider E, Franke R. 2002. Entwicklung eines Erntegerätes für Kulturen von Bergfrauenmantel *(Alchemilla alpina agg.)*. Zeitschrift für Arznei- und Gewürzpflanzen; 7(4): 203-207.

Tesic M, Topalov S, Martinov M. 1987. Machine for chamomile flowers harvesting. Yugoslav Patent. No 20897: 11 p.

Triquart, E. 1997a. Mechanisierte Ernte von Hagebutten. Taspo; 12: 52-53.

Triquart, E. 1997b. Mechanisierte Ernte kleinfrüchtiger Wildfruchtearten. In 1. internationale Wildfruchttagung. Institute für Gartenbauwissenschaften, Fachbereich Technik in Garenbau, Humboldt–Universität zu Berlin, Berlin, Proceeding: 103-111.

Zimmer, S., Müller, J. 2003. Literatursammlung und–auswertung zur Erntetechnologie von Arznei–und Gewürzpflanzen. Fachagentur Nachwachsende Rohstoffe e.V., Gülzow.

Chapter 3

Drying

Milan Martinov
Serdar Öztekin
Joachim Müller

INTRODUCTION

Water is a significant component of biological materials. The physical and chemical properties of crops are determined by their moisture content. The first step in many postharvest operations is the removal of water, i.e., drying. Drying is basically defined as the decreasing of crop moisture content, aimed at preventing enzymatic and microbial activity, and consequently preserving the product for extended shelf life. Drying results in reduction of the weight and volume of the crop with positive consequences for transport and storage.

Drying must meet the following requirements:

- Moisture content has to be brought down to be at an equilibrium level that is defined for certain relative air humidity and temperature. This is defined as storage condition by standards or client.
- Minimum quality reduction in terms of active ingredients, color, flavor, and aroma.
- Microbial count must be below the prescribed limits. No chemical additives may be used.

The moisture content of material is commonly defined on a wet or a dry basis. The *wet-basis moisture content* (X_w, kg water per kg wet material) is defined as follows:

Medicinal and Aromatic Crops

doi:10.1300/5895_03

$$X_W = \frac{M_W}{M_W + M_{DM}}$$

where M_w is the mass of moisture and M_{DM} is the mass of dry matter. It is expressed in percent. The wet-basis moisture content is commonly used in the commercial trade of agricultural products. The *dry-basis moisture content* X is normally expressed as a ratio (kg water per kg dry matter):

$$X = \frac{M_W}{M_{DM}}$$

The dry basis of moisture content is used in research. The relationship between X_w and X is:

$$X = \frac{X_W}{1 - X_W}$$

An equilibrium state is reached when the rate of moisture desorption from the product equals the rate of moisture adsorption from the surrounding air. The moisture content of the product at this stage is known as the equilibrium moisture content (EMC). The relative humidity of the surrounding atmosphere at the same condition is defined as equilibrium relative humidity (ERH) at the particular temperature. A plot of EMC at a given temperature versus the dry basis moisture content of material is expressed as the sorption isotherm. An isotherm obtained by exposing wet material to air of increasing humidity is termed the *adsorption isotherm;* that obtained by exposing the material to air of decreasing humidity is known as the *desorption isotherm* (Mujumdar, 1997). Sorption isotherms are a characteristic of particular materials and have to be assessed by experiments. Figure 3.1 shows exemplarily sorption isotherms for peppermint.

Yeasts, molds, and bacteria develop at a relative humidity above 70 percent. Furthermore, enzyme activity accelerates at relative humidity above 60 percent. Therefore, the final moisture content of the material should be the equilibrium moisture content that corresponds with a relative humidity of 60 percent at the target storage tempera-

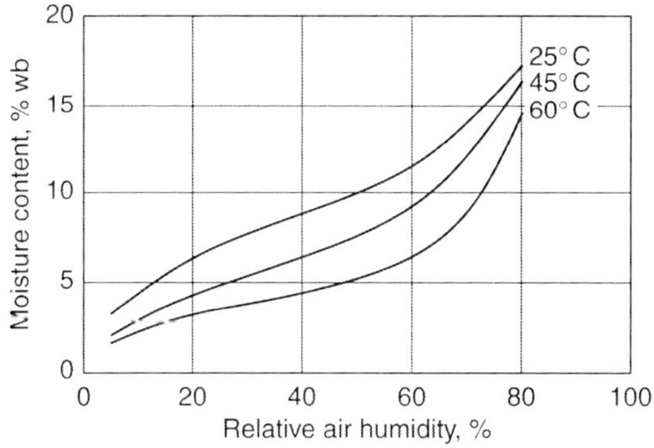

FIGURE 3.1. Sorption isotherms of peppermint leaves (Müller, 2004).

ture, for example, 20°C (Heiss and Eichner, 1990). Over-drying, that is, drying below the above defined final moisture content, is uneconomic, because enhanced drying time reduces dryer capacity and increases energy requirement. Furthermore, the product quality may be degraded.

To evaluate the drying process and dryers, different parameters are used. One of the most important drying parameters is the specific energy consumption that is defined as a MJ (Mega Joulc) of heat energy per kg of water removed. Depending on crop and drying temperature energy consumption ranges from 3.5 to 5.0 $MJkg^{-1}$ in most grains (Sokhansanj, 1997). Considering that the theoretical value is 2.25 $kJkg^{-1}$ for free water, these values show that to extract moisture from grains takes around two times as much energy. Part of this additional requirement is used to heat the grain, part of it to free the bound moisture in the grain, and part is energy needed for transportation of steam to the plant particle surface. The remaining energy is wasted in the exhaust air. The maximum level of drying temperature depends on the crop and the desired quality aspects. The time of exposure to high temperature determines also the resulting quality.

Another important factor to evaluate drying is dryer throughput-capacity. The simplest definition of capacity (kg/h) is annual production divided by total operating time. Throughput may be defined either on a wet or a dry basis. This is average capacity, but real capacity depends on material characteristics and drying air characteristics, i.e., temperature, flow rate, and circulation.

The other important definition for dryer is the *retention time* that is calculated by dividing the mass of dried product by the kg per hour throughput.

Material Safety and Quality

The drying process causes both desirable and undesirable changes to the product. The following changes in product could take place (Bruin and Luyben, 1980):

- Microbial changes and insects population.
- Enzymatic reactions.
- Chemical reactions.
- Physical and structural changes and aroma retention (color and taste, reconstitution properties, volumetric changes).

From the safety and quality point of view postharvest processes play a crucial role to inactivate bacterial growth and to prevent fungal decay in product. It is generally known that the microbial population already exists at the field stage of many crops. This problem is more complicated for harvesting in humid conditions. Microorganisms from the soil could reach the upper parts of the plant in rainy weather. According to Thiede and Beckmann (2004) the amount of natural contamination on MAP could reach even up to 10^7 cfu/g (**c**olony **f**orming **u**nit). The limits of microbial count are already defined in different national and international documents such as European Pharmacopoeia, DAB (German Medicine Book), GAP (Good Agricultural Practices), and the like. In European Pharmacopoeia, the limits for microbial counts are given referring to different applications. Some limits could be obtained from Anonymous (2002).

Increases in contamination after harvest could be a consequence of the transport duration from field to farm, inappropriate storing conditions before and after drying, squeezing during cutting of undesired plant parts before drying, lower drying temperatures, and uneven drying of different parts in batches. The influence of transport and drying method on microbial count of *Hypericum perforatum* L. has already been studied by Graf et al. (2002). The results show that the microbial count could also increase during the transport from field to farm. The drying method influences the microbial count as well. If the drying temperature is over 45°C and moisture content of the material is reduced to under about 25 percent, the development of microbial count is suppressed. The goal is to raise the temperature and lower the moisture content to these levels as soon as possible. This can be achieved in a band dryer because of the low material thickness on bands and the continuous development of process with controlled parameters of drying air in certain sections. In comparison to the band dryer, the spoilage risk in a batch dryer is much higher because of the uneven drying of upper and lower layers. The dried material temperature course in the band dryer and lower and upper layers of the batch dryer could be obtained from Graf et al. (2002).

Any delay in drying could cause rapid spreading of contamination in crop which means, for the grower, either rejection of the product in the market or taking over the sterilization costs. For sterilization there are some thermal applications, such as heated steam and microwaves, that could be used alone or in combination. Heated steam application has a significant influence on reducing microbial count due to the high heat transfer on the product surface, depending on the vapor condensation (Heindl, 2005). In practice, steam generators could be easily coupled in the dryer. The main disadvantage of this technique is the negative influence of a temperature that could reach up to 100°C by atmospheric pressure. Due to this high temperature the cell structure could be destroyed and the cell juices spilt, especially in leafy plants. The resulting influence of these is the collapsing crop mass and mat forming that should be loosened before drying to allow air penetration into the crop. Other disadvantages of heated steam application are the remoistening of a predried crop and the increasing drying time that causes an increase in energy consumption. This process could also have a negative influence on the essential oil content of crops, because heated steam application could actually act as steam distillation. Essential oil losses by steam sterilization are especially important for MAP having gland hairs that are concentrated on the surface of the plant tissue. For example, essential oil losses of steam sterilized chamomile could reach 10 to 20 percent (Kabelitz, 1996). For marjoram, essential oil loss was found to be 25 percent after 3 minutes' steam application where reducing microbial count was 2 log cycles (Hartulistiyoso, 1999).

Other than heated steam application microwaves also provide the opportunity to decrease microbial count. For microwave applications the monitoring and controlling of process parameters such as surface temperature play a crucial role in preserving active ingredients in MAP. To limit essential oil losses to under 30 percent, surface temperature should not be increased beyond a certain temperature, for example 70°C for marjoram (*Majorana hortensis* Moench) (Hartulistiyoso, 1999). By microwave sterilization of marjoram having a moisture content of 80 percent in wet basis the total aerobic mesophyllic bacteria count can be reduced by 1.5 log cycles. The duration of application in this study was 5 min.

The efficiency of microwave sterilization is especially high in a crop with moisture content in the range of 40 to 65 percent (Heindl, 2002). Using microwaves within this range has some advantages, such as acceleration of drying and increasing drying capacity. Under the crop moisture of 15 percent, application of microwaves has some disadvantages. As electricity is usually more expensive than other energy sources, using microwave energy should be limited to moisture ranges with high efficiency (Heindl, 2005).

In practice, steam and microwave techniques could be applied before, during, and after drying. For steam sterilization before drying the crop should be fed in thin layers to the treatment chamber and exposed to steam in a homogeneous layer. After steam application the mat on the band should be loosened by appropriate tools such as a loosing cylinder and the like. It is then forwarded to the dryer. The specific energy demand is about 0.35 kW or 0.5 kg steam per kg of dry material (Heindl, 2005). An example of steam sterilization in a dryer is the five bands dryer where the steam is applied in the extended part of the fourth and fifth band (Heindl, 2005). Due to remoistening of the crop the dryer capacity could reduce in the range of 5 to 10 percent. Specific energy demand in this application is about 0.18 kW or 0.25 kg steam per kg of dry material. For steam sterilization after the dryer there is no need to dry the product again to the final moisture content, which allows increasing the dryer capacity. After sterilization in the special chamber the crop should be dried until the safe moisture stage for storage. Heindl (2005) suggests steam sterilization before drying. In 8 min of steam application for *Cynara scolymus* L. the microbial count could be reduced by 4 log cycles.

But steam application could change the active ingredients, which might pose a problem in getting official permission for using them as medicinal drugs.

Principles of Medicinal and Aromatic Plants Drying

Natural Drying

Natural drying is the simplest method to dry. This is usually acheived for manually collected spontaneous MAP in the sunlight or in a well-ventilated space out of direct sunlight. Crops are dried either in thin layers on trays or in branches. To dry whole branches a proper number of stems are gathered together and are tied into bunches (Figure 3.2). They are hung upside down in a warm and shady place. If the natural ventilation is not sufficient, using a small fan is desirable.

Drying on trays or nets is the method used most often for short-stemmed MAP or for individual leaves or flowers (Figure 3.3 and Figure 3.4). Smaller drying trays fashioned from 2 × 2 cm lumber and screening are placed one on top of the other. By placing spacers at the bottom of each tray, the trays can be stacked in such a way as to allow

FIGURE 3.2. Natural drying in bunches. *Source:* Photo by M. Martinov.

FIGURE 3.3. Natural drying of flowers of *Rosa damescena* Miller on trays. *Source:* Photo by S. Öztekin.

FIGURE 3.4. Natural drying on net. *Source:* Photo by M. Martinov.

good air circulation and overflow, and to take up a minimum of space. The material should be spread in an even layer and should be frequently stirred or moved during the first drying period. This will increase the drying rate and improve the quality of the finished product. Trays should be loaded with no more than about 6 kg of fresh material per m^2. After drying, the completed trays are stored in a clean place until the next time they are needed.

Hot Air Drying

Traditional drying methods, such as drying in the shade or in the sun, have many drawbacks due to the inability to handle the large capacity of mechanical harvesters and to achieve the high quality standards required for MAP. High ambient air temperature and relative air humidity during the harvesting season promote insect and mold development in harvested crops. Furthermore, intensive solar radiation adversely affects quality, causing losses in essential oils or color changes in dried crops. Thus, traditional natural drying in the sun or in the shade does not meet the required standards or consumers demands. To overcome these problems hot air-convective drying is widely used.

Heat energy transfers between a solid and a fluid when there is a temperature difference between them. This is known as *convection heat transfer*. If the temperature of the surrounding air is raised, by some kind of air heater, it can be used as a drying agent. The hot air increases the temperature of solid MAP and causes evaporation of water and other effects. Depending on the surrounding air conditions and temperature difference relative air humidity drops, and air has the potential to absorb water evaporated from the material. This is material transfer from solid MAP to fluid air. Air movement-flow, can be induced by buoyancy, due to the lower density of heated air. This is called *natural convection*. If the movement is caused by ventilators this is *forced convection*. Heated air can flow through a material layer or around it. Nearby convection also causes conduction. This is heat transfer inside solids. The importance of conduction rises with the dimensions of the material. Hot air drying is the most common drying principle today, that is why it will be, together with dryers, presented thoroughly in the following sections.

Microwave Drying

The conversion of microwave-electromagnetic field-energy into heat is explained by two phenomena:

- Water molecules are dipoles and begin to rotate in the rapidly changing electric field. When molecules rotate in a field that changes polarity at a MHz frequency, heat is evolved because of friction forces between the molecules.
- Charge drift under the action of field (ionic conduction). When the ions drift, due to an electric field, they collide with other molecules in a billiard ball fashion and heat is evolved because of friction (Nijhuis et al., 1998).

Microwave drying relies on the principle that energy is absorbed by a wet material when it is placed in a high frequency electric field. This energy is principally absorbed by the water present in the material, causing temperature rise and water evaporation; unfortunately, the electric field is not uniform for higher material layer, and also due to other disturbances, for example, material inhomogeneity. Temperature control can be measured only by expensive spectral thermometers, while the temperature of air is not the same as the temperature of the material. Local overheating can cause degradation of the material.

Appropriate monitoring and controlling microwave drying could open a wider perspective to improve product quality, to reduce drying time, and to increase energy efficiency. Microbial sterilization is already discussed in the previous section.

With microwave assistance, and in combination with convection drying, drying time at hot air dryers can be reduced from 50 to 90 percent depending on the species, duration of treatment, and the amount of energy applied to the crop (Heindl and Müller, 2002). Drying time reduction could be attributed to very high energy concentration in the crop, which means the microwave assisted dryer could be substantially smaller than the conventional hot air dryer for the same capacity. Space requirement is important for installing any system in the building. In progressive drying stages the efficiency of the hot air drying system is greatly reduced, because heat should be transferred from the already dried outer layer to the inner part of the crop. In the case of microwave application, energy is transferred to the moist part of the crop. Thorough drying of a predried product could continue with high efficiency. Microwave energy could also be used in the beginning of drying, and a high drying rate could be obtained in the first drying stage.

Assuming optimal monitoring and controlling of the system are provided, microwave drying could lead to better preservation of active ingredients due to faster drying. The location of active ingredients in parts of plants plays an important role for preservation of quality. MAP contain active ingredients whether in gland cells or hairs at the leaf surface (e.g., *Rosmarinus officinalis* L.), or in the inner part of the capillary system (e.g., *Petroselinum crispum* L.). Plants having active ingredients on the leaf surface are very sensitive to high energy concentration that causes considerable essential oil losses due to cracking and destroying of oil cells. Plants containing active ingredients in the capillary system and dried by microwave have better retention compared to hot air dried variants (Heindl and Müller, 2002). Koller et al. (1998) suggest short application time and low energy input to prevent loss of essential oil. Especially important here is the product temperature. By proper controlling of temperature, microwave assisted hot air drying systems lead to better color retention and lower essential oil losses in *Majorana hortensis* Moench (Rusli, 1998). A high drying rate restricts enzymatic reactions and color does not change. The essential oil content of a microwave assisted hot air dried chamomile sample was found to be 38 percent higher than the pure hot air dried sample (Heindl and Müller, 2002). In addition, the drying time of microwave-hot air combination was 25 percent shorter than the other. Similar results were also obtained for peppermint, hops, and St. John's wort. Energy saving by proper microwave application could reach up to 67 percent in comparison with conventional hot air drying.

Further issues of microwave drying should be considered:

- Using microwaves for MAP should be studied for each special case and plant. For each plant there is optimal moisture content at which microwave energy could be applied efficiently.
- Before microwave application, a metal detector should be used to prevent fire risk of fluorescent metal parts.

- Structure of material, cut plant with high density or whole plant with lower density, plants with or without stem, and stem/leaf ratio influence microwave energy absorption. Example: energy efficiency of microwave application at 2450 MHz is much higher for *Thymus vulgaris* L. having high loading density compared to lower density (Funebo and Skjöldebrand, 2000).
- The geometry of the plant is also important. Browning due to high energy absorption at different points of the plant could cause severe quality reduction. Therefore the cutting and spreading of crop on the band should be as homogeneous as possible (Heindl and Müller, 2002).
- Microwave drying combined with convection drying leads to reduction of microbial count.
- For microwave drying electricity is used. Due to the nature of electricity generation 1 kWh is calculated to be not 3.6 MJ, but 10.8 to 12 MJ. The price of electric energy per J or kWh is consequently higher, with a higher environmental impact as well.

Freeze Drying

Freeze drying is useful for high value, unstable, heat sensitive products such as some MAP. It is achieved by freezing the wet substance and causing the ice to sublime directly to vapor by exposing it to a low partial pressure of water vapor below about 600 Pa (Snowman, 1997). The low partial pressure of water vapor is most conveniently ensured by carrying out the process in a vacuum. Freeze drying takes place in three distinct stages. The first is *prefreezing* to a temperature where the material is completely solid. This can be significantly below 0°C. The second stage is *sublimation* or *primary drying* in which all the free ice is removed. The third and final stage is *desorption* or *secondary drying* when bound water is removed.

Snowman (1997) summarizes the advantages and disadvantages of freeze drying. The advantages are as follows:

- Minimum damage to heat sensitive products.
- Creation of a porous, friable structure.
- Speed and completeness of rehydration.
- The ability to sterile filtered liquids just before dispensing.

The disadvantages are as follows:

- High investment costs of equipment (about three times those of other methods).
- High energy costs (also about three times those of other methods).
- Long process time (typically 16-20 h per drying cycle).
- Possible damage to products due to change in pH and tonicity when solutes concentrate as pure water freezes into ice.

To obtain optimal drying conditions, while maintaining quality, it is necessary to study the freeze drying characteristics of products individually. Tambunan et al. (2001) studied freeze drying of ginger *(Zingiber officinale)* and Javanese pepper *(Piper retrofractum Vahl)* that were mashed into a paste to obtain the dried product in powder form. Freezing was accomplished on a contact plate freezer with a plate temperature of about −42°C to obtain a final temperature of the product of about −32.8°C within 1 h. According to Tambunan et al. (2001) the drying time is influenced more by chamber pressure and freezing rate than by the surface temperature of the product. Higher chamber pressures and faster freezing rates tend to shorten the initial (primary) drying time but lengthen the secondary drying time. The quality of the freeze-dried product was assessed as slightly lower than the quality of the raw material, but higher than when oven dried at 35 to 40°C, and met the related standard

Freeze drying process needs a high amount of energy. Apart from the high investment costs, this is the reason that freeze drying is not economical. It is used only for highest value, price, and material.

HOT AIR DRYING PARAMETERS

The first aim of MAP drying is to preserve material and enable safe storage. This also includes the saving of active ingredients.

Influence of Drying Temperature on Active Ingredients

Gland hairs are epidermal attachments called trichomes. Gland hairs are often formed by just one cell though sometimes several cells are involved. After harvest the surface area of gland hairs reduces parallel to cell dying, where water could be lost drastically. Drying encourages moisture loss from the whole tissue, including gland hairs. Drying temperature is the most important parameter to save the active ingredients of volatile oil in gland cells, which are very sensitive to temperature increase. Drying temperatures from 30°C to 50°C are evaluated as suitable for many MAP species (Buschbeck et al., 1967; Wichtel, 1970; Heeger, 1989; Müller, 1992; Soysal and Öztekin, 1998). The suggested temperatures for MAP depend on the active ingredients and change in the following ranges:

- For glycosides till 100°C
- For slimiest drugs 65°C
- For essential oils till 45°C (Heindl and Müller, 1997).

Generally high temperatures influence essential oil quality and quantity in MAP not only during drying; reduction in active ingredients continues during the storage period as well. Similar results are confirmed by different references (Blazek and Kucera, 1952; Koller, 1987; Shilcher, 1987). According to Jud (cited in Müller, 1992) the maximum drying temperature for *Salvia officinalis* is declared as 30°C. By increasing the drying temperature from 30°C to 55°C, the drying time reduces by 90 percent. But because of this temperature change, essential oil losses increase by 15 percent and the drug color turns from green to gray. To compensate for the negative influence of temperature increase on essential oils, Bendl et al. (1988) have been using the freeze drying method for *S. officinalis*. The quantity of essential oil in the freeze-dried drug was two times higher than the crop dried at 40°C. According to Müller (1992), the quantity of essential oil in leaves of fresh *S. officinalis* differs from 1.4 to 2.5 percent of dry matter. Till 60°C the oil quantity of fresh material is not changed as long as a final moisture of 11 percent is not exceeded. Over 60°C the oil loss increases proportional to the temperature increase. It reaches 30 percent at 90°C. On the other hand, essential oil quantity in *Chamomile recucita* (L.) flowers is not influenced by drying temperature. It changes between 15 and 25 percent, independent of drying (Müller, 1992).

The influence of crop moisture content on essential oil losses for a defined drying temperature was studied by Müller (1992). For drying

S. officinalis at the temperature of 60°C oil losses occur first at a moisture content of less than 11 percent (Figure 3.5). To an equilibrium moisture content of 2.5 percent losses increase very rapidly till 50 percent. At 90°C drying temperature 10 percent of essential oil is already lost at 40 percent moisture content. The reason for oil losses at high temperatures are not the bursts on oil glands as expected. Müller (1992) indicates shrinkages on gland cell as the main reason for oil losses. To explain the complex diffusion mechanism of essential oil more research is required. But all results show that Salvia can be dried at a temperature of 50°C without any losses. Even the extended drying time at that temperature has no negative impact on essential oil.

Drying temperature usually has an influence on the temperature sensible components of essential oil (Table 3.1). For example, easily volatilized components of essential oil from *S. officinalis,* such as α, β-pinen, camphen, and myrcen reduce with increasing temperature (Müller, 1992).

Influence of Drying Temperature on Color

Drying temperature has important effects on the color of MAP. The changes in the color parameters of fresh and dried peppermint

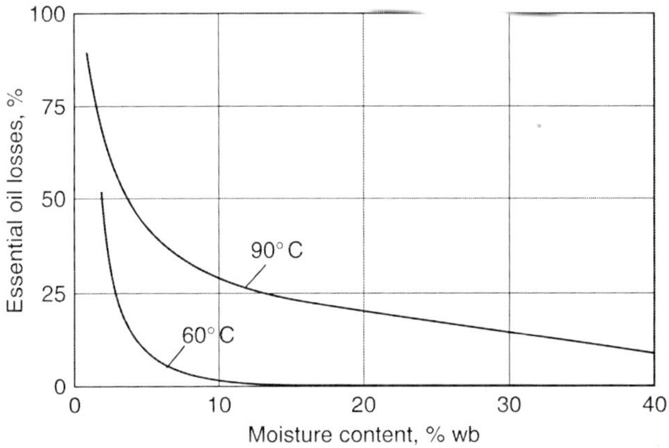

FIGURE 3.5. Influence of crop moisture content on essential oil losses of *Salvia officinalis. Source:* Müller, 1992.

TABLE 3.1. Influence of temperature on composition of essential oil in *S. officinalis.*

Component	Fresh (mg/ml)	30°C (mg/ml)	90°C (mg/ml)
α-pinen	23	50	14
Camphen	30	57	19
β-pinen	23	26	0
Myrcen	13	10	0
Cineol	107	62	69
α-Thujon	601	347	326
β-Thujon	83	99	45
Campher	117	158	181
Borneol	14	28	19
β-Caryophyllen	15	23	31
α-Humulen	25	34	38
β-Humulen	4	3	0

Source: Müller, 1992.

(*Mentha piperita* L.) as affected by drying air temperature were given in Soysal (2000). Statistical tests showed that the color parameters of the fresh peppermint differed significantly from the dried products exposed at different drying air temperatures (Table 3.2). Higher temperatures could cause dramatic color deteriorations and burning of the product. Therefore, temperatures below 50°C may be suggested as optimum drying air temperatures to obtain quality products in terms of color.

Influence of Drying Temperature on Specific Drying Energy

To compensate for the influence of suggested low drying temperatures for MAP, which reduce drying speed, the dimension of the drying system should be increased. This also means an increase of investment costs. The second parameter that influences operational costs is high energy requirement. To withdraw the high amount of moisture from fresh crops, a large amount of energy is needed. For example, due to the high initial moisture content of 83 percent wb

TABLE 3.2. Effect of various drying air temperatures on the color of peppermint (*Mentha piperita* L.).

Color parameters	Drying air temperature (A) Fresh	(B) 30°C	(C) 45°C	(D) 60°C	(E) 70°C	LSD
L (%)	−40.40±3.74a	37.97±2.81b	36.54±1.37b	31.31±1.54c	30.87±2.77c	2.105543
a (%)	−11.05±0.91a	−8.16±0.75b	−6.63±1.17c	−3.58±0.31d	−2.17±0.97e	0.736884
b (%)	14.08±2.32a	13.86±0.98a-b	12.72±1.59b	10.23±1.05c	10.06±1.53c	1.288070
α (°)	128.44±3.02a	120.48±1.32b	117.47±3.18c	109.40±2.04d	102.05±5.04e	2.689195
C	17.92±2.32a	16.09±1.18b	14.37±1.79c	10.84±1.02d	10.33±1.57d	1.358426

Source: Soysal, 2000. Reprinted with permission.

[a, b, c, d, e]Different letters in the same row indicate a significant difference ($P < 0.01$).

(or 4.93 db), 815 kg water have to be evaporated per 1,000 kg of fresh parsley leaves to attain the final moisture content of 9 percent wb. Energy demand is defined as specific drying energy. The specific drying energy–energy input for drying, depends on the material dried, material initial and final moisture content, drying process, and type of dryer. It is calculated as primary energy of fuels used per kg of evaporated water or dried or fresh material. It is mostly expressed per kg of dried material.

Drying temperature influences specific drying energy greatly. According to the experiment of Müller (1992) done in a laboratory dryer, the results show a high influence of drying temperature on specific drying energy of *S. officinalis*. It was 13 MJ/kg at drying temperature 30°C and increased to 42 MJ/kg at 40°C (Figure 3.6). From the energy point of view *S. officinalis* should be dried in that case either at 30°C or over 50°C. On the other hand drying temperature should be considered also from the energy input and costs point of view.

Influence of Air Relative Humidity on Drying

The air relative humidity is a measure of its capacity to absorb water, i.e., steam, from dried material by convective drying processes. During the drying process two changes of drying air occur: the air tem-

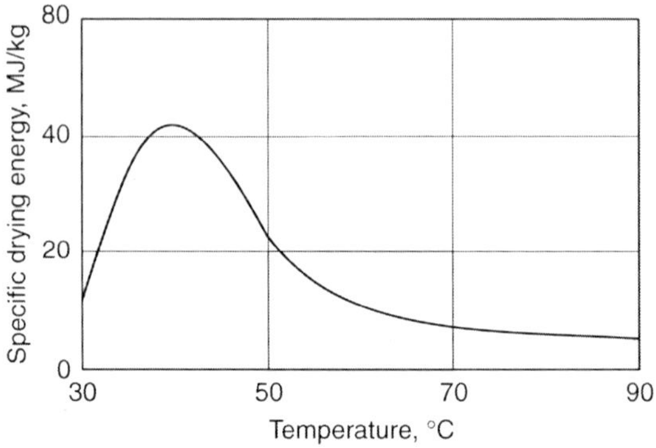

FIGURE 3.6. Specific drying energy per kg of dry sage (Dew point temperature:13°C, air velocity: 0.2 m/s. *Source:* Müller, 1992.

perature drops, due to convective heat exchange, and water in a steam form is taken from the plant material. Both cause the increase of air relative humidity. Water absorption is a dynamic process and its intensity is mostly influenced by air relative humidity.

The intensity is also influenced by the amount of evaporated water and its position. The absorption is more intensive if the amount of water is higher and if it is located on the material's surface. This is the case for the first drying phase of wet material. The high air relative humidity has the consequence of prolonging the drying process. For producers, it is important to know the limit values of drying air relative humidity for economic drying. For the beginning of the drying process of wet material, moisture content over 40 percent, the limit value of drying air relative humidity is 60 percent or more. For the drying of material of moisture content in the range from 40 to about 20 percent it is about 40 percent, and for lower moisture content it is about 20 percent. These data are the results of a survey of practitioner's experience. This issue needs further investigation. There is very limited information about the influence of relative humidity on quality parameters of different MAP species. For sage and chamomile, for example, no significant losses of essential oil have been obtained in relative humidity under 40 percent (Müller, 1992). The available knowledge shows that the monitoring and controlling of relative humidity in hot air drying could be evaluated as a problem of variable costs. At low levels of relative humidity where hot air drying occurs, quality changes could be ignored. To find out the influence of relative humidity on components of active ingredients more detailed research should be carried out for different MAP.

Influence of Air Velocity on Drying

The influence of air velocity on convective drying is related to the volume of air moved through the dryer and the pressure drop. These parameters define the ventilator power and other characteristics. Air velocity influences convective temperature exchange between drying air and the material. On the other hand, just as the air temperature rise is obtained by heat transfer between air and a heat exchanger, the required air volume is a function of air velocity and surface area of the heat exchanger. Of course, in open mode, the thermal power should be adequate to be able to bring the temperature difference to the air volume. The air velocity does not differ according to the dried plant part or species. But, due to pressure drop in progressive drying stages, air velocity could be reduced to save energy either by closing sack openings or RPM reduction of engine. In practice, air velocity for MAP drying in batch dryers is selected between 0.08 and 0.2 ms^{-1}. For band dryers it is usually up to 0.75 m/s. For example, air velocities in band 1/2/3/4/5 from bottom to the top are 0.75/0.60/0.45/0.3/0.15 m/s, respectively.

Over this range a considerable increase in electric consumption is expected without adequate acceleration of the drying process.

HOT AIR DRYERS

Type of Dryers

Dryers can be divided into three major groups according to the material flow during the drying process:

1. Batch dryers
2. Semi-continuous dryers
3. Continuous dryers

In batch dryers the whole batch of loaded material is dried in one charge. After drying, the charge dryer is unloaded and then fresh crop is filled as a second batch. In some cases, producers add some material in the batch after certain drying time, but the drying course is, in this case, not substantially changed. This is the most common and widely used type of dryer for MAP. The design is rather simple, and purchasing and construction costs are relatively low in comparison to other dryer types. Thermal power range is from 50 to 400 kW. Facilities with thermal power under this range are designed for hobby production. Energy consumption is between 5 and 15 MJ per kg evaporated water. Efficiency depends on design, applied drying process, possibilities of process control, plant material, condition of the plant material, and so on. Loading and unloading of batch dryers is only occasionally mechanized, and these operations require a lot of labor. This is commonly used by small and medium producers, especially in developing countries. Batch dryers are mostly suitable for bulky herbal materials, but they can be used for flowers, roots/bulbs, and grains drying, with an appropriate grate and drying layer height. Ideally, it is profitable if it is used more than three months per year.

The most significant problem concerning batch dryers is the distribution of drying air. Due to the flow characteristics of air the highest air velocity is obtained at the end of the box, and that is why the boxes should be as long as possible. Some air diverters can contribute better uniformity of the air flow. Uniformity of plant material distribution is of great importance. The holes and places with much lower material density cause the formation of tunnels and "free" flow of drying air through the material without any drying effect. The consequences are insufficient drying of material and energy losses. Another problem is local compression of material and creation of nests. The material in this location will not be properly dried. Due to the above mentioned effects, the moisture content of the material is not uniform. With one or even two turnings over of the whole charge, the drying process can be improved, but it needs additional labor. An example of a simple on-floor batch dryer is presented in Figure 3.7. The ventilator (2) pushes the hot air into the main duct (3). The smaller lateral ducts (4) are connected to the main duct. The lateral ducts are open from bottom–trapeze roof form, and raised from the floor, creating distance-openings for air flow. The space within the main duct is filled

FIGURE 3.7. On-floor batch dryer. 1–air heater, 2–ventilator, 3–main duct, 4–lateral ducts distant from the floor level. *Source:* Photo by M. Martinov.

with the plant material up to their upper edge. In the on-floor batch dryer, bulky herbal material, like peppermint, can be dried. The drying of cut material, leaves, flowers, fruits, and the like is possible only with additional netting laid on the lateral ducts (4), but lower efficiency is expected. The dryer can be easily and quickly assembled on any solid level ground, e.g., a concrete surface, whereby only the main duct has to be sealed. Disassembly can also be done quickly, and the same space could be used as storage or for other purposes.

The great disadvantage of this type of dryer is the high energy input. Even if the main duct is insulated, the drying agent (air) is always going out to the environment. The flow of air through the material is not uniform, and some plants are in direct contact with the hot metal sheets. This type of dryer was frequently in use in the 1960s and early 1970s. The high energy prices today dissolve the main advantage of this dryer type—its low investment costs—within one drying season. The plant material is exposed to insects and other pollutants during the drying process, which can cause contamination. From this point of view, this type of dryer is inconvenient as well.

Commonly used batch dryers are of a box type with a net or air permeable textile floor. The box is situated in a container or building. The walls and roof of the container or building should be well insulated and coated with hygienic paint or other material inside. To enable drying air input and distribution, the floor can be elevated or channels can be dug beneath the box. One example of this kind of batch dryer is presented in Figure 3.8. Due to better distribution of drying air, the length of the box is usually three to four times bigger than the width. For a bigger dryer, multi boxes are used, as in Figure 3.8. Uniform air distribution is achieved by a longitudinal duct with reducing cross-section surface (4), and adjustable sliders for each box (5). Big box gates (8) enable easy filling and emptying of boxes. While drying material with high moisture content, they are in the beginning left open to enable the outflow of high humid air. Steered slider (10) closes and opens wall opening (9), which enables air to flow from drying room to the heater room. If it is closed the surrounding air is drawn through the vents of the heater room door (11). If it is open the air circulates from the drying room.

Cabinet dryers with drawers, shown in Figure 3.9, are used to dry smaller amounts of plant material. This is a very popular solution for

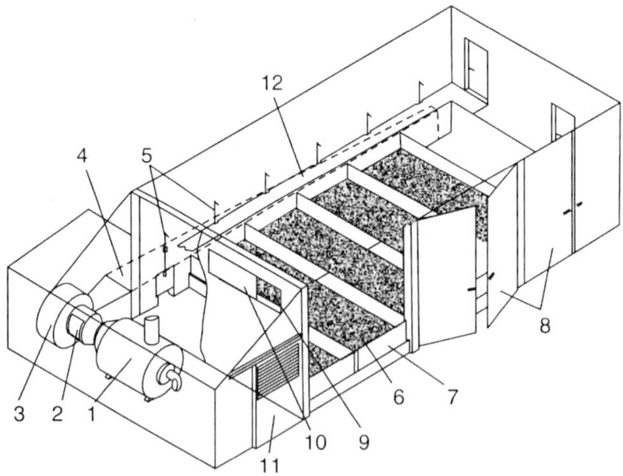

FIGURE 3.8. Batch dryer with six boxes with the total surface of 90 m^2 and 350 kW thermal power. 1–air heater, 2–device for blending with ambient air, 3–ventilator, 4–distribution duct with inconstant cross-section, 5–sliders for air flow adjustment, 6–box with grate floor, 7–front sliding wall of box, 8–gates, 9–wall opening, 10–steered slider, 11–heater room door with vents, 12–inspection road. *Source:* Martinov, 2004.

hobby production and is used for testing in R&D and demonstration institutions. Here the drying air flows through the plant material placed on the net of drawers. Cabinet dryers are suitable for the drying of cut/sliced material, leaves, flowers, and grains.

Besides the facilities described, there are various other batch dryers designed for special purposes. One example is the grain dryer with a rotary agitator, Figure 3.10. The dryer is characterized by a rotary agitator, which moves along the box and turns the plant material over, here oil pumpkin seeds. The rotary agitator can also be used for the box discharging at the end of the process.

The basic characteristic of semi-continuous dryers is intermittent filling and emptying of the plant material. The drying process is almost continuous. Due to the movement of the plant material, adequate parameters for each drying phase can be applied. The intermittence of filling, i.e., emptying, depends on material load, air heaters characteristics, thermal power, and air flow. The thermal power range is from 100 kW to 1 MW, and energy consumption is from 4.5 to 8 MJ per kg evaporated water. Almost all dryers are suitable for sliced herbs, small fruits, grains, flowers, and the like. Most of them are also suitable for mushrooms, sliced fruits, and vegetables. Investments for

FIGURE 3.9. Combined solar-electric cabinet dryer used for the testing of the quality of medicinal plants. *Source:* Photo by M. Martinov.

FIGURE 3.10. Special box dryer with traversing rotary agitator for pumpkin seed drying. *Source:* Photo by M. Martinov.

this type of dryer are higher than for the batch ones with the same capacity, but the duration of the redemption period could in some cases be even shorter due to lower energy costs. Profitable use of a semi-continuous dryer requires its engagement for three to four months per year.

Figure 3.11 shows a multilevel, i.e., multitier dryer commonly used for hops and pepper drying, but also for other MAP. It has been developed from the traditional multilevel dryers used in central Europe. The fresh material is filled on the upper tier, and it falls down by the tipping of the tiers segments (6). The material flow is opposite to the drying air flow—it is "falling" down. The upper tiers are made of tipping, perforated sections. The lowest tier is a permeable band conveyor (5), intermittent activated, and used for discharge. For some solutions all tiers are made of band conveyors.

The material on the upper tiers has a higher moisture content and needs higher drying capacity of drying air. This is why the adjustable flap in the air duct (3) leads a certain amount of the hot drying air directly to the upper zones (9). There is another adjustable flap in the exhaust air duct (7) that diverges part of the air back to the heater through the circulating air duct (8). This enables reduction of the

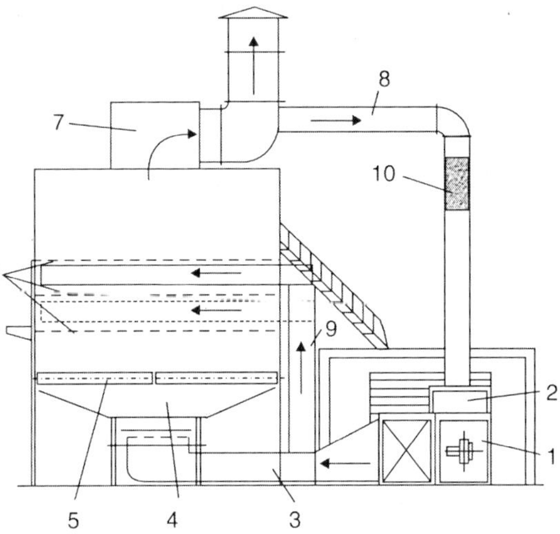

FIGURE. 3.11. Multitier–falling bed dryer. 1–air heater, 2–fresh air ventilator, 3–drying air duct, 4–air distributor, 5–discharge band conveyor, 6–tipping tiers, 7–exhaust air ventilator, 8–circulating air duct, 9–additional hot air (for upper tiers) supply duct, 10–filter for circulating air duct (prevents that light plant parts come to air heater and cause fire). *Source:* Heindl, 2001. Reprinted with permission.

energy input, but it is efficient only if the exhaust air has lower moisture content than the ambient air. This happens during the second half of the intermittence period, when the moisture content of the material in the upper tiers has been reduced. Charging and discharging of this dryer type can be relatively easily mechanized, and the labor reduced. Homogenous distribution of the plant material to prevent energy losses is also one of the main issues here. A mobile multitier dryer tipping sections and its tipping mechanism are shown in Figure 3.12.

The widely used tunnel dryer with truck-racks is presented in Figure 3.13. Here the plant material arranged on racks is placed in a truck. That is why this type of dryer is also called a shelf dryer. Racks are made of metal or wood frames with net bottoms. All the front and back sides of the racks frame are so designed as to enable air flow. This is an overflow type of dryer—air does not stream through the material, but over and under thin layers. The tunnel can be designed for two or more trucks. One additional truck is always used for charging and discharging of material. The dryer is suitable for small or

(a)

(b)

FIGURE 3.12. (a) Perforated tipping sections of tier, (b) tipping mechanism. Source: Photo by M. Martinov.

sliced fruits, cut material, sliced mushrooms, and grains. Charging is in most of the cases manual, which causes high labor costs, and discharge is by gravity emptying. Due to the flow of drying air, material in the first truck will be dried first. After reaching the needed level of final moisture content, the first truck is pulled out and replaced by one with fresh material. After a certain period, a more or less constant

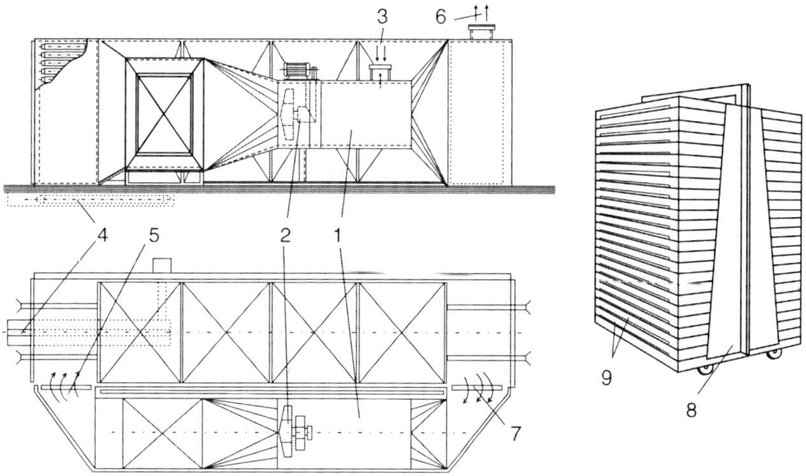

FIGURE 3.13. Tunnel dryer with side positioned air heater. 1–air heater, 2–ventilator, 3–fresh air inlet, 4–truck pulling mechanism, 5–drying air inlet, 6–exhaust air outlet, 7–flap for air circulation, 8–truck, 9–racks.

intermittence period of truck is achieved. Manual pulling or pushing is possible for dryers with up to three trucks. If the number of trucks is higher, mechanical pulling (4 in Figure 3.13) would be needed. Partial circulation of drying air is provided by the positioning of the flap (7). This enables reduction of energy inputs. Air circulation is not effective immediately after charging of a new truck, because of the intensive evaporation of fresh material.

The dryers with constant charging and discharging, i.e., flow, of plant material are known as continuous dryers. The most common are band dryers, with stainless steel or plastic permeable bands. There are designs with one to five bands. A one band dryer offers the best possibilities for drying process control, but it is also the most expensive. The drying process, after starting the first charge, runs continuously. The different drying phases are performed on certain sections of a band or on different bands. Material charging and discharging also run continuously. Due to the known positions of the drying phases different parameters of drying air can be applied at different bands, that is, band sections. This enables better drying quality and energy consumption reduction. Thermal power range of these dryers is from 0.7 to 3 MW, and energy consumption is from 4.5 to 5.5 MJ per kg evaporated water. Due to high investment costs the dryer would be profitable if it is used five to six months per year. This also means that vegetables and fruits should be dried, especially the kinds that could be inexpensively stored until the time of drying. Typical for extended drying time are bulb and root plants, for example, onion, celery, and the like. Band dryers are suitable for cut or sliced material, not for bulky herbs, small fruits, and grains; for these materials some special equipment is needed. Bulky herbal materials must be prepared by cutting and separating of stalks and stems; this contributes to energy savings because the stems are not dried, but it creates additional investments for the cutting and separating. Other materials, fruits, and vegetables need different preparations for drying in the band dryer, i.e., additional equipment.

All band dryers have the possibility of band speed regulation, via an infinitely variable mechanical transmission or a frequency converter. Band speed should be so adjusted as to get the highest capacity and most complete drying to the defined moisture content of the material (equilibrium moisture content). This is the task of experienced operators and it can be supported by adequate process-control devices, for example, on-line measuring of moisture content.

Band dryers have sections, by which different temperatures and air flows could be adjusted. Dryers with higher thermal power can be equipped with two or more air heaters, which makes easier adjustment of drying air characteristics for certain sections, that is, drying phase.

Homogenous distribution of plant material over band surface is, like for other dryer types, of essential importance. Here, only mechanized charging and discharging are applied. Drying of sticky material can also cause difficulties, such as ineffective cleaning of bands and blockage of openings. The side view of the five bands dryer is shown in Figure 3.14. Swinging band elevators oscillate left-right and provide, together with feeding rollers, a uniform layer of material on the upper drying band. Rotating brushes (4) clean the upper bands after the reload of material to the lower ones. Ventilators (7) transport exhaust air through the duct, whereby certain parts can be circulated to the air heaters.

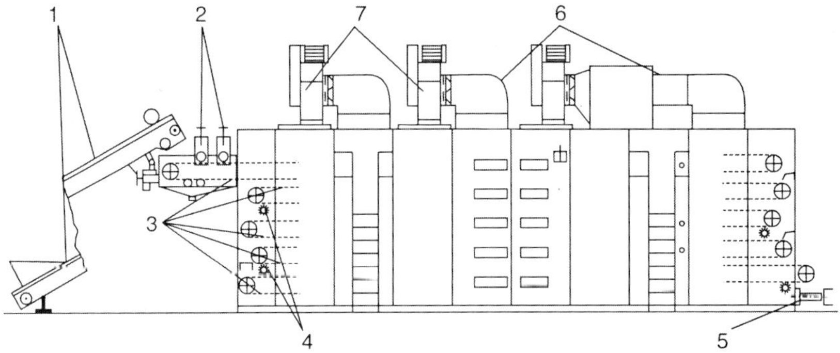

FIGURE 3.14. Band dryer. 1–swinging charging band elevator, 2–dosing rollers, 3–bands, 4–rotating brushes for dry cleaning, 5–discharging band conveyor, 6–exhaust air ducts, 7–ventilators and air circulating ducts with flaps. *Source:* Heindl, 2001. Reprinted with permission.

Figure 3.15 shows the cross section of this dryer. In this dryer type, three air heaters supply different bands. The first one supplies only the first, upper band, because a higher temperature of heating air is necessary. The plant material with the highest moisture content is on the first band. Higher temperature cannot cause damages to active ingredients because the energy is transformed into latent heat for intensive water evaporation, i.e., the temperature of the plant material is kept under the limited value. Drying air distribution flaps (1) share the total air amount to the bands. Flaps (6) and (7) also regulate circulation of air. Circulating air is aimed primarily at the lower bands, because the drying air in these drying phases is not saturated in one pass. The amount of circulated air is controlled manually or automatically.

The most sophisticated method of reduction of heat losses in exhaust air is based on heat recovery in the air to air heat exchanger. This heat exchanger, with a big contact surface made of stainless steel or aluminum alloy, enables recovery of up to 50 percent of exhaust gas energy by heating of input air. In this case, no regulation of air circulation is needed, nor is additional manual or automatic equipment. High investment in such device can be paid back in less than one drying season if the dryer is optimally used.

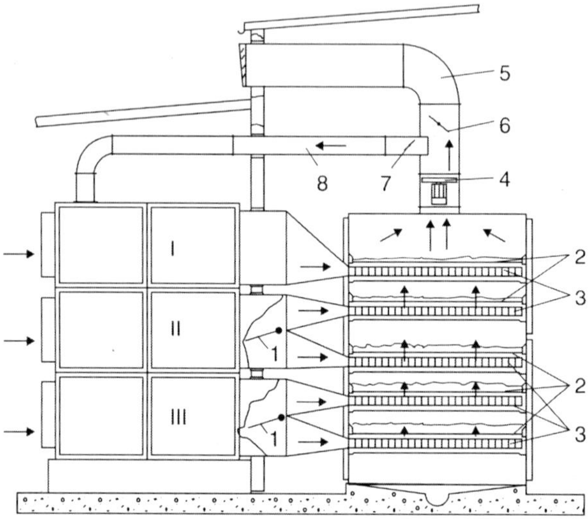

FIGURE 3.15. Cross section of the band dryer with five bands and three air heaters. I-III air heaters: 1–hot air distribution flaps, 2–bands, 3–perforated tubes for air distribution across band width, 4–exhaust air ventilator, 5–exhaust air duct, 6–exhaust air flap, 7–circulating air flap, 8–circulating air duct. *Source:* Heindl, 2001. Reprinted with permission.

Dryer Selection, Monitoring, and Control

Selection of Dryer

The selection of a dryer is one of the most important decisions when setting up a production line for MAP. The first task is to choose the appropriate drying capacity for one or more dryers. This is an interactive decision, because sometimes the growing plant, or the usage of drying capacities of other producers should be taken into consideration. In addition, an economic analysis should be done, a drying price calculated, and a redemption plan made. For higher investments, e.g., a band dryer, the production of MAP should be, in most cases, combined with the production of vegetables and fruits, and drying service should be provided for other growers. In this case, some additional predrying processing lines are needed, which involves further investments.

The selection of the dryer type is also influenced by the form of products. If bulky herbs are to be dried, without predrying procedures of cutting and stem separation, tunnel truck-racks and band dryers are not suitable. On the other hand, the drying of sliced or cut material cannot be successfully performed in box batch dryers. For the drying of collected spontaneous flora and servicing of several growers, a band dryer is not adequate. It cannot be easily adjusted for heterogeneous material. The same applies to drying of MAP grown in remote fields. Long distances cause increased costs and material degradation due to long transport. Generally, the period between harvest and drying may not exceed 3 h. This time could depend on weather conditions. High temperature and high air humidity reduce the above mentioned limit.

Today what is applied is almost exclusively "indirect" drying. This means that combustion gases are not used directly for drying but for heating of the drying agent (air) via heat exchanger. MAP drying by exhaust gases is not permitted.

Fuel selection is an important issue, primarily when it concerns the price of energy. The price of natural gas is usually the lowest, followed by LPG (light petrol gas) and heavy heating oil. The price of light heating oil is usually the highest. The price relation depends on local conditions, and energy and tax policy. Each fuel needs an adequate burner; the lowest price is for a light heating oil burner. Combustion efficiency is the highest and the emission of pollutants the lowest for natural gas and LPG. Solid fuels, coal, can be used as a fuel as well, but a special air heater/heat exchanger is needed. The use of renewable energy sources is discussed in a separate chapter.

Process Monitoring and Control

Monitoring and control of the drying process is of crucial importance for drying quality and efficiency. What is most important is to know and control the drying air temperature, because it has the biggest influences on the drying process and material quality. Automatic control of the drying air temperature, which should have limited fluctuation from the set value, for example ±2°C, has to be implemented even in very simple dryers. This temperature has to be regulated over the control of combustion process and the blending with fresh air. It can be done manually, but today it is typical, and not expensive, to use automatic adjustment of heating air temperature. The temperature of the exhaust air is the next important element, and it can be obtained by a simple thermometer. The difference in temperatures before and after passing the dried material gives information about the drying intensity. If the difference is higher, the drying process is more intensive.

Relative drying air humidity, after passing the material, is valuable information. Psychrometers and hydrometers cost more than thermometers and are less accurate. The accuracy of a good instrument in a certain range, for example, 20 to 60 percent, is ±3 percent. Most common is the use of the differential psychrometer, where air humidity is calculated from the difference between dry and wet temperature. The data on drying air humidity provide information about air saturation, and it is the background for the decision on when to switch from circulating to open mode by batch dryers. This switch can be performed manually or automatically. In the case of manual control the operator has to be present all the time, to ensure proper and the most economic drying process.

An example of an automatically controlled box batch dryer is presented in Figure 3.16 Drying air thermometer (3) is connected with the burner of the air heater and enables desirable temperature with fluctuation of ±2-3°C. Another thermometer, limit thermometer (10), is a safety thermometer that stops fuel and electricity supply if a certain maximal temperature value is reached. This can occur in the case of fire. This thermometer also activates alert signals.

Open air flow mode is performed if the side flaps of the air heater (9) are open and the upper wall flap (15) is closed. The drying air, after passing the plant material (15), flows out through overpressure vents (14) situated on the gate (13). Two servo motors can close the side flaps and open the upper flap, which provides circulated air flow mode. Drying air comes back to the air heater. The servo motors are activated according to the drying program via the process computer (4). The mode selection is performed in accordance with the relative humidity of the drying air. This humidity is measured with a differential psychrometer, "dry and wet" temperature thermometers (5 and 6). For example, the mode can be changed from circulating to open if the relative humidity reaches 35 percent. After switching to the open

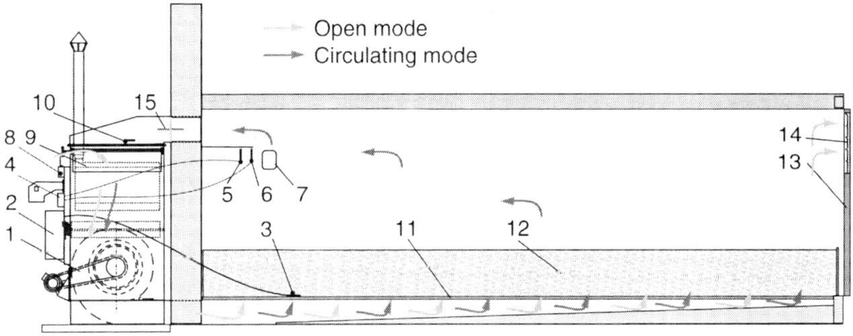

FIGURE 3.16. Cross section of batch dryer with automatic selection of open and circulation operation mode. 1–air heater with ventilator, 2–command box with temperature limiter 3–thermometer for drying air, 4–process computer, 5–thermometer for "dry temperature," 6–thermometer for "wet temperature," 7–water supply vessel, 8–servo motor, 9–side flaps, 10–limit temperature-safety thermometer, 11–grate floor, 12–dried material, 13–insulated gate, 14–over pressure vents, 15–upper flap with servo motor. *Source:* Martinov, 2004.

mode the ambient air will be sucked into the air heater. Due to temperature increase the relative humidity will be reduced significantly. The duration of the open air flow mode can be set up and should enable replacement of the former saturated drying air in the whole dryer.

All band dryers, especially new ones, have automatic control of the drying process. This includes changing of dry air parameters, steering of air circulation and speed of bands according to the achieved material moisture content at the final point. Air flow rises during the drying process, especially in batch dryers, because of material pressure drop reduction. Reduction of ventilator flow, using a frequency converter, contributes to lower electricity and fuel consumption.

Material Moisture Content Monitoring

Monitoring of material moisture content, especially at the final stage, is an important issue of every drying process. If the material has not reached equilibrium moisture content, which enables safe storage, microbial count rises and quality degradation occurs. If the material is excessively dried, the drying process is prolonged and unnecessary energy input is made. This causes losses of essential oils and other active ingredients in many MAP. It is also important to know the material moisture content during the drying process. This is the background for some decisions concerning drying parameters. An example is the adjustment of band speed. Many small and medium producers rely on their experience—feeling. An experienced producer is able in many cases and for many different MAP to assess correctly the final product's moisture content. This is not the case for fresh or partly dried material. Standardized laboratory moisture content methods, aimed at research purposes, are very time consuming. A typical method is drying in ovens for 24 or even 72 h. This is not practically applicable. Some tests with a microwave oven using up to 600 W have shown that the measurement error is within 2 percent. The measuring takes 20 to 30 min. This method could be good for quick monitoring of moisture content by batch dryers. Moisture content measuring instruments based on capacitive—dielectric—properties method are widely in use for grains, especially for cereals and maize. These are used for MAP grain products, pumpkin seeds for example. Producers already supply buyers with adequate tables to read results, but for some materials calibration has to be done. For band dryers the continuous in-line measuring of moisture content is needed. Spectral, infra red, that is, near infra red (NIR) instruments for moisture content measuring are used today for many materials. The principle is based on measuring the reflected light at specified frequencies. These instruments can measure moisture content only on material surface, and they are less suitable for operating in dirty environments. Substantial calibration is needed for every particular plant material. Changes in the color of the material can also cause a measuring error. Calibrations are already applied for numerous materials but only for limited moisture content ranges (Heindl and Hendl, 1998). The water molecules oscillate, i.e., rotate, in a magnetic field, as has been already explained in the section related to microwave drying. The consequence is the heating of the water itself and the surrounding material. As the amount of water in certain materials is higher, the reflection of microwaves is lower because the energy is consumed for molecule oscillation-rotation. The frequency of the radio frequency generator can be changed and the reflected waves measured. The peaks of the curve representing amplitude–frequency relation are used to define moisture content. Frequencies in the range from 0.4 to 10 GHz are used. This is a low-energy instrument, with the power of only 10 mW. The intensity of measuring depends on the properties of the material. Due to new developments in digital techniques and data processing, up to 25 readings per second are possible. Measuring is also possible in harsh environments (Anonymous, 2005). The application of microwave moisture content sensors needs substantial calibration. These sensors are already used in the industry, but further improvement and price reduction is expected in the near future. Both NIR and microwave sensors deliver digital information that can be easily used for automatic control of the drying process.

IMPLEMENTATION OF RENEWABLE ENERGY SOURCES

Due to worldwide shortage of fossil energy as well as environmental problems, e.g., CO_2 imbalance, the use of renewable energy sources (RES) is an important issue of research, development, and practice. Unfortunately, implementation of RES is not an easily achievable goal. It is especially difficult to achieve it profitably and in an environmental friendly manner without additional investments or input of labor. Many institutions and programs are dealing with this problem, trying to reach adequate usage of RES. However, this will remain a serious objective for future R&D activities and challenge for practice.

Solar Energy

Solar energy is traditionally used for on-field or in-yard drying of MAP, by direct exposure, or covered by plastic foils. In all cases, the drying time is long and the conditions for microbial development are favorable. Weather conditions, insects, and air pollution affect the on-field and open air-drying. Due to this, solar drying does not result in safe and quality products in many cases. The regulations of S/Q are getting stricter, and this kind of drying has prospects only for some MAP, for example, grains, Figure 3.17. In addition, direct solar

FIGURE 3.17. On-field drying of milk thistle. *Source:* Martinov, 2004.

radiation can cause degradation of active ingredients of some products, such as chamomile.

Direct solar drying in plastic covered spaces is widely practiced in many developing countries, especially in the tropics and subtropics. Covering products causes a rise in temperature and drying acceleration (Figure 3.18), and protects from insects and other pollutions. It is common, in some regions, to use greenhouses during the non-growing season for drying purposes, but the capacity of such dryers is rather low, which should be taken into consideration while production planning.

One modification of a direct solar dryer is the tunnel dryer, Figure 3.19, whereby the air is circulated with the use of a fan (Mühlbauer et al., 2001). The solar heater is very simple; it consists of an elevated platform with insulated bottom and sidewalls. The first half of the length is used as the air heater and the second half is filled with the material. Both sections are covered with a transparent foil.

Nevertheless, the design of this solar dryer is very simple, it gives good results and is used in practice. Typical use is for drying of grapes, figs, and apricots, but it can also be successfully used for drying of MAP, primarily fruits and grains. The problems exist in regions with lower solar radiation and during instable periods of the

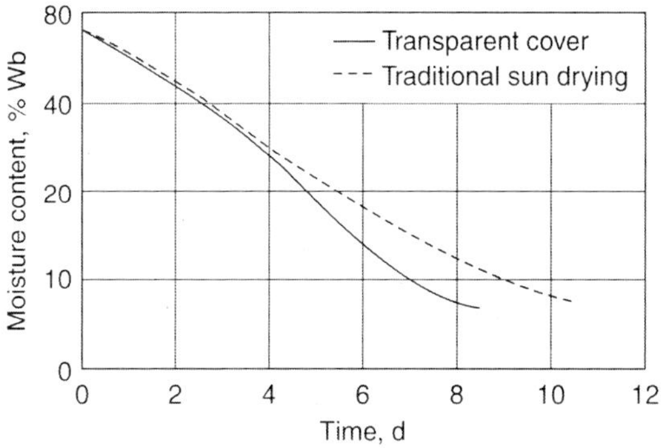

FIGURE 3.18. Comparison of drying time by traditional open-air sun drying and drying with transparent cover. *Source:* Müller, 2004.

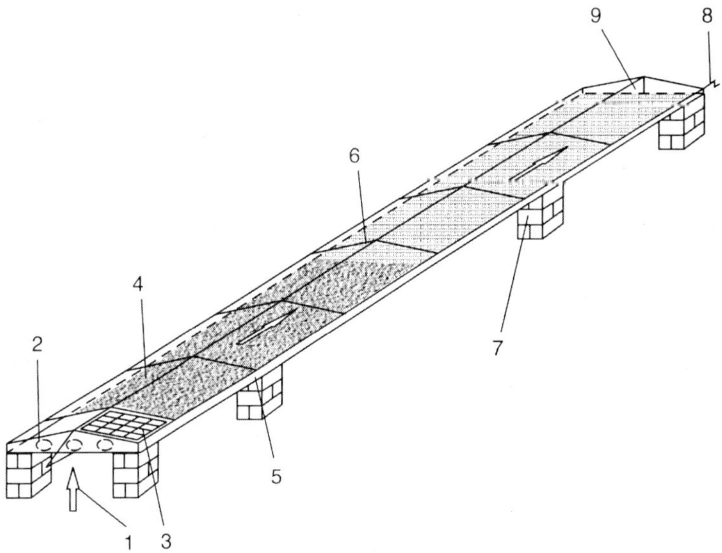

FIGURE 3.19. Tunnel solar dryer. 1–air inlet, 2–fan, 3–solar generator, 4–solar air heater, 5–metal frame, 6–drying tunnel, 7–block structure, 8–rolling bar, 9–air outlet.

year. The heating capacity is low per occupied surface, which can cause high initial investment costs in some regions or on some locations. For profitable use it is of importance to select appropriate working parameters and to select adequate inexpensive materials for construction.

In *indirect solar dryers* the material is not directly, or through foil, exposed to solar radiation. The solar heater raises the temperature of ambient air, reducing its relative humidity, and this is used as a drying agent.

The simple type of indirect solar dryers with natural convection, known as cabinet or shelf dryers, are popular for hobby drying of small amount of MAP. Forced dryers, with a ventilator, can be used by small and medium MAP producers. The roof of a dwelling or warehouse is often used as a solar heater. An example is solar assisted warm air drying that can be applied for band dryers, whereby inlet air is preheated by passing through, for example, roof solar heater (Heindl, 2000).

A successful solution for solar drying is a modified plastic green house, Figure 3.20. Here the south-oriented side of the roof is extended to the ground level, to increase the surface exposed to the sun. The inclination depends on the location. The inclination in central Europe is about 30°, which enables most efficient heating in the summer

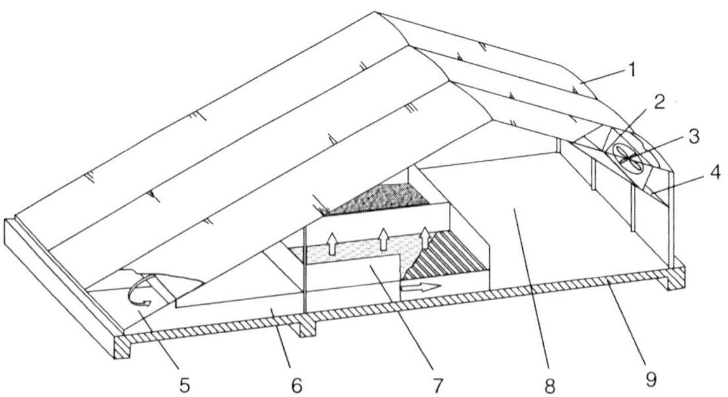

FIGURE 3.20. Green house based indirect solar dryer with forced convection. 1–transparent cover, 2–absorber, 3–fan, 4–recirculation flap, 5–expansion chamber, 6–air duct, 7–batch dryer, 8–passage, 9–basement. *Source:* Müller, 1992.

time. The heater consists of an absorber and transparent foil, which are placed in the slots of two separate guiding laths. The absorber foil can be removed and the facility used as a green house out of the drying season.

The material is in the drying boxes with a grid floor. The circulation flap is controlled and enables setting up of maximum agent temperature. This also enables regulation of the drying process by selecting the open and circulation mode, according to the agent relative humidity in the dryer. The fossil fuel hot air generator can be added to enable drying during the night.

Solid Biomass

Using of biomass as a fuel should be an interesting solution for many producers, especially if they can use crop residues and process trash from their own production. Like other plant materials, residues with 15 percent moisture content have net heating value of about 15 MJ/kg. The combustion technique of solid biomass depends on the form of the material, that is, bales, briquettes, chips, and the like. Facilities for bales or chopped straw-like material are difficult to control if an air generator is used. Discontinued combustion causes huge oscillation of air temperature and is not applicable for drying purposes. This can be overcome by using boilers and a heat exchanger for air heating. In this case, what is required is to install a big heat exchanger (thermal energy transfer from about 70 to 50°C) with high pressure ventilator in addition to the boiler. The costs for an additional heat exchanger and high pressure ventilator are comparable with costs for complete liquid or gas hot air generator. The reasonable solution would be to use the same boiler for other purposes, first of all for heating. In this case, the needed thermal power should be similar for both purposes. By using a heat accumulator the nominal heating power for drying may be up to twice that for heating, and operating hours of the boiler may be adequately reduced. To summarize, the using of boilers for solid biomass and combustion of bulk and bales of straw-like materials can be economically feasible only in combination with heating or other energy use with comparable thermal energy needs.

Some other materials and forms of solid biomass are already being used for drying purposes in hot air generators (Kaltschmitt and Hartmann, 2001). Typical are wood chips, residual fruit kernels, and pellets. The combustion process should be continuous, with stepless fuel feeding. In a good facility it is possible to achieve a temperature fluctuation of ±2°C. Such facilities are already used in practice.

Another example is the hot air generator for wood logs with two-step combustion, Figure 3.21. The zones of gasification and combustion are almost disconnected, which enables better control and more efficient combustion. This results in reduction of pollutants in exhaust gases. The combustion can be controlled by combustion airflow, where an under-pressure type is better than an over-pressure one. The problem of such a facility is that it causes breakdown of electricity supply, due to the development of high temperatures in the combustion zones, which are not cooled if the main ventilator is not in operation. Some facilities aimed at pellet combustion are applicable for drying, but the prices of pellets are higher than other solid biomass forms. The emission of pollutants, CO, NO_x, dust, and the like are limited, and these regulations must be followed.

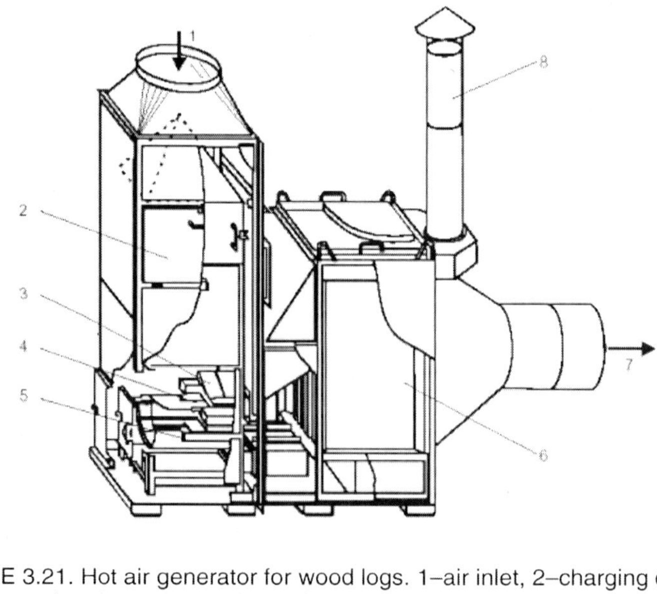

FIGURE 3.21. Hot air generator for wood logs. 1–air inlet, 2–charging door, 3–primary combustion zone, 4–slot, 5–fire clay tube, 6–tube type heat exchanger, 7–hot air outlet, 8–chimney. *Source:* Martinov and Babic, 1994.

Liquid and Gaseous Biomass

Plant oils, native and their esters (PME–plant oil methyl ester, RME–rape oil methyl ester), alcohol-methanol, and biogas are the most important renewable liquid and gaseous fuels. The esters of plant oils are used primarily as fuel for internal combustion diesel engines and their use as a heating fuel is uneconomic. Some burner types could be used for combustion of native plant oils and light heating oil blends, for example, heavy heating oil burners. Light heating oil burners must be adapted for using native vegetable oils, equipped with preheating facility. Widmann et al. (2001) explain the possibilities of using native plant oil as a fuel for heating and drying purposes. Plant oils can be, as pure or blended with light heating oil, used only in special burners with automization and preheating.

The prices of native plant oils are still too high in comparison with other fossil fuels. Therefore, their wider use for drying purposes could be only profitable infrequently, usually with the help of subsidies. The situation is similar for the use of alcohols, that is, methanol. The use of biogas as a fuel is possible with the same burner as for natural gas, but its use is also a question of profitability. Other possibilities of biogas and vegetable oils use are CHP (combined heating and power) facilities. Within the drying season there is no need for heating, and thermal energy could be used for electricity generation. The problems are the same as for the use of solid biomass; the needed drying thermal power should be the same as the residual heat of the CHP facility.

Obstacles and Prerequisites for Usage of RES

High investments for suitable facilities, difficulties in achieving appropriate drying parameters, and the huge surface needed for solar heaters are the most significant barriers for RES implementation. These could be overcome in the case of considerable rise of fossil fuel prices and by adequate energy policy accompanied with subsidies. Nowadays, reduction in the emission of CO_2 and other greenhouse effect gases is set up as a priority in the field of environment protection. Different measures, apart from the implementation of RES, have been and will be introduced at the national and international levels. The following should be considered before making the decision to apply renewable energy sources for the drying of MAP:

1. The potential of planned renewable energy source has to be thoroughly analyzed.
2. The needed capacity of the dryer should be defined with particular needs at certain periods of the year, and compared with the RES potential.
3. Risks and energy supply security should be analyzed.
4. The characteristics of the potential facilities, and the possible demerits, should be studied.
5. Feasibility study, investment sources, and payback have to be carefully considered, as well as any stated subsidies or bonuses.

Solar Energy

Low power of units causes higher investments and occupation of a large surface. The maximal radiation per square meter is about 1,100 W and it changes during the day, just like the effective surface of the heater. The average solar radiation power is, according to this, about 35 percent maximum, with about the same efficiency of a solar heater. It means that the average power is maximum 150 W/m^2 for 12 h in June per day in the north hemisphere. That would be about 1.8 kWh of net energy per day. For a well-designed dryer it means about 1.2 liters of evaporated water. For example, the heating oil dryer with 150 kW net power, occupies a surface of less than 100 m^2 and removes about 2,400 liters of water in 24 h. For the same capacity about 2,000 m^2 solar heater is needed. This surface should be multiplied by two for solar drying at the end of September or if the day is partly cloudy.

Solar energy can be successfully and profitably used if roofs of houses are used for heaters, if a greenhouse type of dryer combined with a hot air generator is used, or if the preheating of air is implemented. Depending on the location and the weather conditions in summer a saving of 0.2 to 0.3 liter of heating oil per m^2 of solar heater per day can be achieved (Heindl, 2003).

Solid Biomass

A potential user should estimate his own potential of generating biomass as well as the prices. If the potential of own sources is not sufficient, the quantity and price of biomass in the vicinity should be analyzed. The prices of hot air generators using biomass should be considered, including additional equipment. The additional input of labor and other merits and demerits of using biomass should be studied. In most cases, the use of solid biomass would not be profitable without governmental support. There are already practice applicable solid biomass hot air generators for fruits kernels, wood chips, and logs. It is important that the facility has a good control over the drying. Bulky biomass, like straw bales, can be successfully used only in combination with a boiler. In that case the needed thermal power for both applications has to be similar.

Liquid and Gaseous Biomass

Use of native plant oils, pure or blended, and biogas is currently not profitable without subsidies. The use of thermal power of CHP plants can be successful and profitable in some cases. Essentially, the thermal capacity of a facility should match the dryer's needs.

REFERENCES

[Anonymous] 2002. Europaeisches Arzneibuch. Amtliche deutsche Ausgabe. 4. Ausgabe. Stuttgat/Eschborn: Deutscher Apotheker Verlag/Govi Verlag.

[Anonymous] 2005. Keam Holdem Industrial RF & Microwave Technology, available from: www.kha.co.nz.

Bendl E, Kroyer G, Washüttl J, Steiner I. 1988. Research on freeze drying of thymus and salvia (In German). *Ernaehrung/Nutrition;* 12: 793-795.

Blazek Z, Kucera M. 1952. The influence of drying methods on active ingredients of chamomile (In German). *Die Pharmazie Bd.;* 7(2): 107-109.

Bruin S, Luyben KChAM. 1980. Drying of food materials: A review of recent developments. In: *Advances in Drying*, Volume 1, Mujumdar AS. Editor. Hemisphere Publishing Corporation in corporation with McGraw-Hill International Book Company.

Buschbeck E, Keiner E, Klinner J. 1967. Physical and thermal properties effecting drying characteristics of peppermint (In German). *Archiv für Landtechnik;* 2: 163-200.

Funebo T, Skjöldebrand C. 2000. *Aroma of Thyme Dehydrated by a Combination of Microwave Energy and Air - Influence of Process Parameters.* SIK. The Institute of Food and Biotechnology. Göteborg, Sweden.

Graf C, Schubert E, Thiele K, Müller J. 2002. *Hypericum perforatum* L: Veraenderung des mikrobiologischen Status waehrend Ernte, Transport und Trocknung. *Z. Arzn. Gew. Pfl;* 7.Jg: 31-37.

Hartulistiyoso E. 1999. Untersuchungen zur Application von Mikrowellenenergie für die Entkeimung von Gewürzen am Beispiel von Majoran. PhD. Universitaet Göttingen.

Heeger EF. 1989. *Handbook for Medicinal and Aromatic Plant Cultivation* (In German, Reprint version of first Edition from 1956). VEB Deutscher Landwirtschaftsverlag, Germany.

Heindl A. 2000. Solare Warmlufttrocknung von Arznei- und Gewürzpflanzen. *Z. Arzn.Gew.Pfl.;* 5(2): 80-88.

Heindl A. 2001. Facilities for drying of medicinal and aromatic plants. *Medicinal Plant Report;* 8(8): 17-22.

Heindl A. 2002. Mikrowellenunterstützte Trocknung von Arznei und Gewürzpflanzen. *Zeitschrift für Arznei&Gewürzpflanzen;* 7(Sonderausgabe): 208-225.

Heindl A. 2003. Probleme und Lösungsmöglichkeiten bei der Aufbereitung und Trocknung von Arznei- und Gewürzpflanzen. *Z. Arzn.Gew.Pfl.;* 8(1): 33-36.

Heindl A. 2005. Möglichkeiten der Keimreduzierung bei Arznei- und Gewürzpflanzen vor, waehrend und nach der Trocknung mittels Dampf- und Mikrowellenapplikation. *Arznei und Gewürzpflanzen;* 10(2): 100-105.

Heindl A, Hendl T. 1998. Kontinuerliche Feuchtemessung zur Regelung von Bandtrocknern. *Z.Arzn.Gew.Pfl;* 3(3): 146-154.

Heindl A, Müller J. 1997. Trocknung von Arznei- und Gewürzpflanzen. *Zeitschrift für Arznei&Gewürzpflanzen;* 2: 90-97.

Heindl A, Müller J. 2002. Mikrowellenunterstützte Trocknung von Arznei- und Gewürzpflanzen. Z. Arzn.Gew.Pfl., special edition concerning Fachtagung für Heil- und Gewürzpflanzen, Bad Neuenahr- Ahrweiler, November 12-15, 2001, 208-225.

Heiss R., Eichner K. 1990. *Haltbarmachen von Lebensmitteln: chemische, physikalische und mikrobiologische Grundlagen der Verfahren.* Springer-Verlag, Berlin.

Kaltschmitt M, Hartmann H. 2001. *Energie aus Biomasse.* Springer-Verlag, Berlin.

Koller WD. 1987. *Problems with the flavour of herbs and spices.* Proceeding of the fifth Int. Flavour Conference, Porto Karras, Chalkidiki, Greece, 123-132.

Koller WD, Raghavan B, Range P. 1998. *Quality of Microwave Dried Medicinal Plants and Herbs.* Federal Research Centre for Nutrition, Alliance product and nutrition research, Karlsruhe, Germany.

Martinov M, Babic M. 1994. Development of hot air generator using wood log as a fuel. *Savremena poljoprivredna tehnika;* 20(4): 184-188.

Mühlbauer W. 1989. Grain Drying. In: *Drying Technique* (In German). Band 3. Springer Verlag, p. 64-93.

Mühlbauer W, Bux M, Ritterbusch S. 2001. *Renewable Energy for Rural Areas.* Teaching materials. University of Hohenheim. Institute for Agricultural Engineering in the Tropics and Subtropics, Stuttgart.

Mujumdar AS. 1997. Drying Fundamentals. (In: *Industrial Drying of Foods,* First Edition, Baker CGJ. editor. Blackie Academic & Professional, London, p. 7-29.

Müller J. 1992. *Trocknung von Arzneipflanzen mit Solar Energy.* Dissertation. Ulmer Verlag. Germany.

Müller, J. 2004. *Drying of MAP.* Teaching material. University Stuttgart-Hohenheim, Institute for Agricultural Engineering, Stuttgart.

Nijhuis HH, Toringa HM, Muresan S, Yuksel D, Leguijt C, Kloek W. 1998. Approaches to improving the quality of dried fruit and vegetables. *Trends in Food Science & Technology;* 9: 13-20. Online at http://www.infodienst-mlr.bwl.de/la/lap/pflqual/analymet/Trocknung_E.doc.

Rusli M. 1998. Versuche zur kombinierten Mikrowellen- und Konvektionstrocknung von pflanzlichen Produkten, dargestellt am Beispiel von Kartoffeln und Majoran. Dissertation. Fakultät für Agrarwissenschaften. Georg-August-Universität Göttingen, Germany.

Shilcher H. 1987. *Chamomile* (In German). Wissenschaftliche Verlaggesellschaft mbH, Stuttgart, Germany.

Snowman JW. 1997. Freeze dryers. In: *Industrial Drying of Foods.* First Edition, Baker CGJ. editor. Blackie Academic & Professional, London, p. 134-155.

Sokhansanj S.1997. Through-flow dryers for agricultural crops. In: *Industrial Drying of Foods.* First Edition, Baker CGJ. editor. Blackie Academic & Professional, London, p. 31-63.

Soysal Y. 2000. *Research on Fundamentals of MAP Drying and Developing a Field Scale Cabinet Dryer* (In Turkish). PhD. Thesis. Cukurova University, Adana, Turkey.

Soysal Y, Öztekin S.1998. Effects of Drying on the Color and Total Chlorophyll of Mentha Piperita. AgEng Oslo 98 International Conference on Agricultural Engineering, 24-27 August, 1998, Oslo-Norway, Report No: pdf/f/98-F-016.pdf (printed on CD).

Tambunan AH, Yudistira FW, Kisdiyani, Hernani. 2001. Freeze drying characteristics of medicinal herbs. *Drying Technology;* 19(2): 325-331.

Thiede S, Beckmann G. 2004. Mikroflora von Arzneipflanzen-Vorkommen und Bedeutung Enterobactericeae. 2. Mitt.: Eigene Untersuchungen. *Zeitschrift für Arznei&Gewürzpflanzen;* 9(3): 130-140.

Wichtel M. 1970. Changes in active ingredients at drug preparation. *Die Pharmazie;* 25(11): 692-698.

Widmann B, Stelzer T, Remmele E, Kaltschmitt M. 2001. Produktion und Nutzung von Pflanzenölkraftstoffen. In: *Energie aus Biomasse.* Kaltschmitt M., H Hartmann, editors, Springer-Verlag, Berlin, p. 537-583.

Chapter 4

Mechanical Processing

Milan Martinov
Miodrag Konstantinovic

INTRODUCTION

The term mechanical processing comprises diverse operations of postharvest treatment of MAP. The aims of mechanical processing are to eliminate undesirable components, refine plant material, and adjust their form according to market demands or to nurture them in the best condition for further processing. Colloquially, the term *dry processing* can also be used here to distinguish differences between mechanical processing and other procedures where water, steam, and solvents are used as agents, for example, distillation and extraction. The term *mechanical processing* is more precise than the term *dry processing*, because this part of the production chain includes, for some plants, washing and water/solvent separation.

Mechanical processing could include a lot of different operations. It is hard to formulate a widely accepted classification valid for all, or almost all, MAP. This chapter and the given list of processing steps should be considered only as a kind of proposal for discussion that enables easier understanding and communication between the people involved in production or R&D activities in this field.

According to the "place" of operation in the whole production chain, postharvest mechanical processing can be divided into:

1. Predrying processing.
2. Postdrying processing.

Medicinal and Aromatic Crops

doi:10.1300/5895_04

Certain producers practice predrying processing, especially the large-scale ones. It includes washing, for some crops, for example, of roots and bulbs, and various preparations for more efficient drying. The most frequent operations will be described in the following subsections.

Postdrying processing is more in use, and it includes the following operations:

1. Plant parts removal—threshing
2. Size reduction
3. Separation—classification
4. Others

PREDRYING PROCESSES

Cleaning–Washing

A very common operation belonging to this production phase is the cleaning of harvested material, especially the roots and bulbs. There is a wide range of procedures and machinery—equipment used. Generally, there are two possibilities:

1. Dry-cleaning
2. Washing

Dry-cleaning primarily removes soil residues from roots and bulbs using specially designed brushes, rotors with metal/rubber fingers, or vibrating sieves. Sometimes the loosening of harvested material is required as the first phase of cleaning, performed to reduce water consumption and to preserve the best soil on the field. This operation is much more efficient if the soil is dry. The example of dry-cleaning is presented in Figure 4.1. A rotating drum with fingers is used to agitate material and loosen the bonds between plant material and soil. After this, a coarse sieve is used. It is even better to have this facility as a part of the harvester or to perform this operation on the field.

Small producers, particularly in the developing countries, use available water sources for washing. Still waters are not as suitable as the flowing ones. Still waters are often contaminated and efficient washing needs moving of the material. In hilly regions, it is the prac-

FIGURE 4.1. Dry-cleaning of harvested roots. *Source:* Reprinted with permission from Müller, 2000.

tice to put roots-bubbles, e.g., valerian, in cages, in water streams. The contamination of used water has to be controlled always. If the production is small, the material can be washed by splashing, whereby grate/screen and water drainage measures are to be used, Figure 4.2.

The other simple possibility is to use common concrete/mortar mixers for this purpose. These are available as masonry tools on many farms. It is, of course, necessary to clean the mixer before use and to change the water to achieve proper washing. In this case, only uncontaminated water is allowed.

For removing soil and other impurities, washing pools could be used. For this purpose, some permitted chemicals should be added to prevent development of microbes. It is used for plant material with specific density lower than water, that is, for materials that float on water. The sediment at the bottom of the pool consists of impurities and should be periodically removed, and the water changed.

Many industrial washers, primarily designed for vegetables, can be used for cleaning of MAP. The user should evaluate the washer according to its efficiency, capacity, energy and water consumption, labor needed, investment, maintenance, and other costs. Every washer

FIGURE 4.2. Improvised washing of small amount of roots using hose and grill-screen.

is designed to achieve the best possible washing and minimal water consumption. Water must be chemically and microbiologically approved. Water characteristics are of crucial importance. If circulation of water is applied it must be properly cleaned, and it is obligatory to fulfill local and international regulations. Circulation of water can be done only if it has been tested; just a few cycles are usually possible. Having one's own well of usable water is a big advantage, because technical water treatment can be very costly. The use of tap water is only seldom right, due to the costs involved and because of both global and community water shortage.

Usage of chemicals is not permitted in many developed countries, except flocculation agents, although they could be very mild and neutral.

Geyer (1999), gave very functional instructions for the washing of vegetables, and this can be used for MAP too. The washing machine types are as follows:

1. Drum type, with open or closed drum
2. With nozzles, tunnel or rotated
3. With brushes
4. Flotation washers

There are also machines that combine two or more basic types, or special machines aimed at one or a few products only. A typical drum washer, suitable for small and medium MAP producers, is shown in Figure 4.3. This is the most significant type of simple design with low water consumption.

A specially designed washer with two drums and two nozzles lines, aimed at pumpkin seed washing, is presented in Figure 4.4. Plant slime and other mainly vegetable impurities are separated here. Drum openings are selected in such a manner as to enable separation of vegetable admixtures.

Depending on market requests, some herbaceous plants or leaves have to be washed. This is usually performed by surface spraying of a thin material layer, so that all plants are intensively treated with sprayed water. It is desirable to remove the surface water from plant material after washing. On the other hand, the drying will consume more energy. The requirement is to do it with low energy consumption and labor input using a simple facility. After the herbs or leaves are washed, a horizontal conveyor with stainless steel chain transports the material. During transport, an intensive airflow generated using ventilators is applied from below. The length of this conveyor is 3 to 6 m.

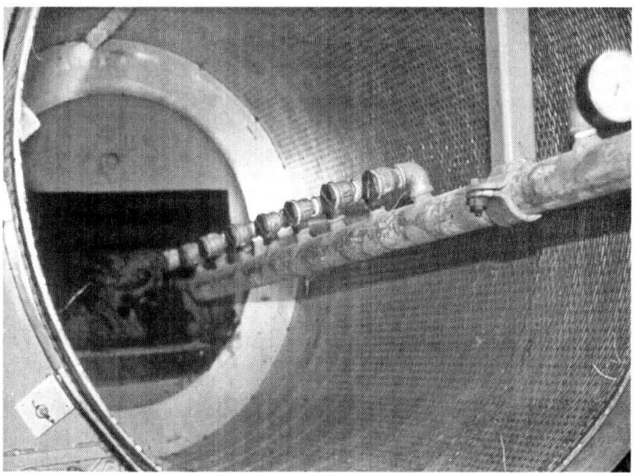

FIGURE 4.3. Drum washer with nozzles, frequently used for MAP.

FIGURE 4.4. Double drum washer with nozzles designed for pumpkin seeds.

Another method for removal of surface water is to use an oscillating conveyor, 2 to 3 m in length, made of stainless steel sieve or wire net. Oscillations force the water to separate from the material and to fall down through the sieve or wire net. For the same capacity and efficiency, this facility is smaller and cheaper than the former one.

For nozzle washers it is important to select appropriate parameters, to generate effective water flow or droplet impulse. This makes washing more effective and reduces consumption of water and material damages (Mulugeta et al., 2003).

The brush type washer is used in combination with water spraying. An example of this washing machine is presented in Figure 4.5. This type of machine can be applied very efficiently for the washing of some root MAP even in small and medium farms. For roots and bulbs with density lower than water, a washer can be used. A scheme for such plants is given in Figure 4.6. The same can be used for some fruits too.

Small and some medium-size producers use simple gravity flotation washers, metal, or plastic boxes with perforated plate over bottom. It takes some time to achieve satişfying soil removal from plant

FIGURE 4.5. Washing machine with brushes for roots.

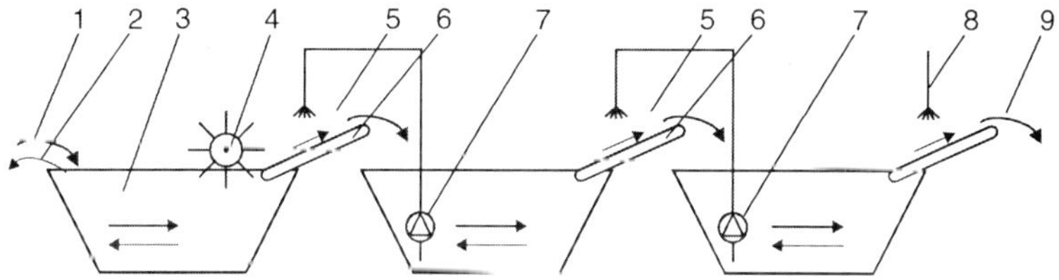

FIGURE 4.6. Flotation washer with three washbasins and counter flow. 1–feeding, 2–dirty water, 3–sparkling basin, 4–paddle roller, 5–cascades, 6–sieve band elevator–surface water removal, 7–pump, 8–final washing with tap water, 9–clean products. *Source:* Geyer, 1999. Reprinted with permission.

material. This procedure cannot be successfully used if the outdoor temperature is high, due to the risk of microbial development.

Predrying Separation and Classification

The goal of predrying procedures is mostly to remove undesirable plant parts: separation, to dry separately different plant parts; classification, or to prepare the plant for the dryer; cutting with/without additional separation and/or classification. These procedures are more typical for bigger producers where band or shelf dryers are used. The simplest operation is manual removal of undesirable parts on the inspection table, or better, on the inspection band. A slow moving band is used, and a couple of persons, in a comfortable position, remove weeds, stones, and other admixtures. If the band dryer is used, with the thermal power from 0.6 to 2 MW, the cutting and separation of stalks and stems are the typical predrying procedures. The processing line consists of three major parts: feeding/dosing conveyor, special cutter, and air flow separator, Figure 4.7. The harvested material is unloaded into the conveyor hopper (1). Dosing unit consists of band elevator (2) and adjustable rotated scraper with fingers (3). By choosing the band scraper distance, the amount of feeding material is adjusted. Material falls to the cutter. The cutter feeding band consists of a lower (4) and upper (5) rubber band, with speed regulation. The position and load of upper band are adjustable. That enables soft treatment of feeding material and the required compression for successful cutting. Usually, the disc-type cutter is used, with two or three special knives. The knives are shaped with spiral edges to perform

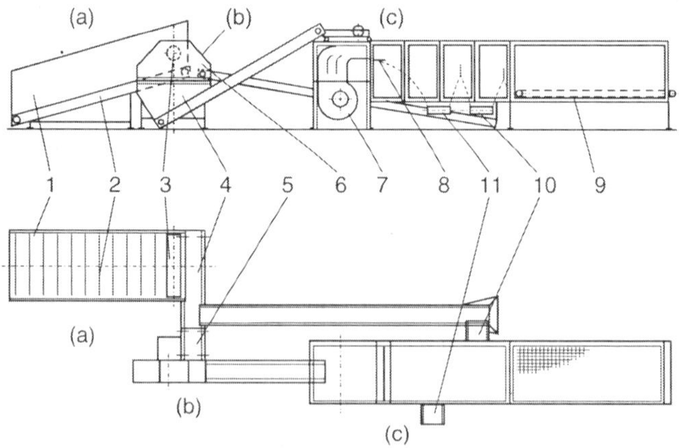

FIGURE 4.7. Facility for cutting and leaves-stalks predrying separation of herbaceous MAP. (a) feeding: 1–hopper, 2–band elevator, 3–dozing scraper, (b) cutting: 4–lower feeding band, 5–upper adjustable feeding band, 6–disc with knives, (c) air flow separating: 7–ventilator, 8–airflow direction adjustment, 9–outlet for leaves–band conveyor to the dryer, 10–outlet for leaves with stems–backward to the cutter, 11–outlet for stems. *Source:* Heindl, 2003. Reprinted with permission.

gradual cutting. After cutting, the chopped material is conveyed to the airflow separator. It consists of a ventilator (7) that generates airflow. The intensity of airflow is adjustable, using RPM regulation or air inlet regulation by side openings. There are also devices for airflow direction adjustment (8). The material falls into the airflow and due to difference in airflow resistance, the material is separated into three groups: leaves, leaves with residual stems, and stems/stalks parts. Leaves are directed toward the dryer or washer (9), leaves with residual stems return back to cutter (10), and chopped stems/stalks to the disposal.

The usual share of stalks and stems in harvested herbaceous MAP is approximately 50 percent. Using the procedure described these plant parts are removed before drying. The consequences are lower energy input and higher output of dryer. Also, one of the merits is that this operation actually provides processing without postdrying threshing. The demerit is a need for higher investments and reduction of essential oil content due to mechanical treatment of fresh, sensitive plant material. This operation is, from the point of view of band or shelf dryer users, almost inevitable due to the needed uniform covering of dryer active surface.

The attempt to develop similar equipment for small and medium producers was not successful (Martinov et al., 1992). Three different types of cutters: adapted forage harvester drum type cutter, specially designed cutter with horizontally positioned knives, and vertical feeding and hammer mill, have been tested in combination with a simple airflow separator using a combine harvester cleaning fan. Only the forage harvester cutter gave acceptable cutting results. But the essential oil losses were very high due to very intensive influence of rubbing. The content of essential oil of uncut peppermint was 4.0 percent and of fine cut material was 2.2 percent. This was the reason to abort further research.

As the "byproduct" of this experiment, a significant reduction of drying time was recorded due to the length of the material cutting. By maintaining the cutting length of 100 to 120 mm the essential oil content after drying was 3.1 percent, and the drying time was reduced by more than 30 percent. This effect has been explained by the cutting of stalks and the destruction of its fiber structure. This caused the shortening of water migration paths and thus the much more effective drying. It must be once again emphasized that most fresh herbaceous plants are very sensitive to any mechanical damage of the active ingredient holders.

Another typical example of predrying separation is used for the processing of chamomile. The harvesting-integrated separation has been partially described in Chapter 2. If there is no separation during harvesting, it is recommended to do it afterwards, in the farmyard and before drying. Separate drying of first and second class harvested material enables better quality and higher dryer output. Predrying processing/conditioning of some plants, primarily separation-classification, can be realized by different procedures and machines. A typical example is predrying separation-classification of chamomile. For this purpose a plane sieve, with minimal 1 × 2 m surface can be used. The sieve openings depend on the variety and are usually 18 to 20 mm. This is not very effective due to material flow without intensive turning. The advantage is that the same machine could be used for other separation and classification jobs. The other solution is the use of single or double drum separators with cylindrical sieves, which are used in other fields as well. Inclination of the rotating drum enables material flow and overturning. Material is intensively moved and considerably better separation is enabled. In addition, a special rolled tube separator for removing of small plant parts and other impurities can be mounted. The separator with a rolled tube is presented in Figure 4.8.

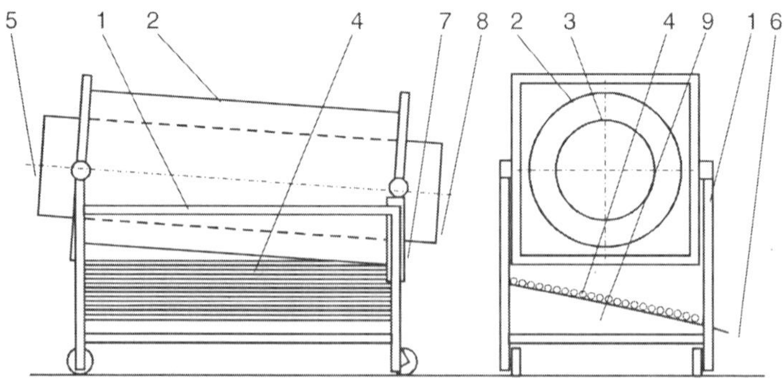

FIGURE 4.8. Drum sieve for fresh chamomile. 1–frame, 2–outer drum, 3–central drum, 4–rolled tube separator, 5–material inlet, 6–outlet of first and second class material, 7–outlet of third class material, 8–long stalks/stems material and weeds, 9–impurities and small plants admixture. *Source:* Lacak, 1998.

The openings on the central drum sieve are usually 20 mm, on the external drum sieve 25 mm, and the length of the sieve is 3 m. The diameter of the tubes is about 20 mm with a distance of 5 mm. The separator enables very good separation of first and second class. The axial outlet of the central drum are plants with long stems and weeds. They can be stored easily due to lower density and a possibility of active ventilation. That enables better use of the dryer capacity, which in most of the cases is too low for full production. The outlet for longer stalks and stems is shown in Figure 4.9. Some other processes, for example, peeling, cutting, kernel removal, epidermis perforating and the like, could be a part of material pre-drying preparation.

Some of the necessary predrying separation procedures have not been solved until now. A typical example is the separation of stigmas and stamens of saffron flowers. If it were possible to separate the stigmas and stamens before drying, a better quality of final product could be achieved. An innovative solution for this operation is presented in Figure 4.10.

The harvested flowers are thrown into the airflow generated by the ventilator (4). Stigmas and stamens are easily separated from the

FIGURE 4.9. Central drum output—long stalks/stems blossoms.

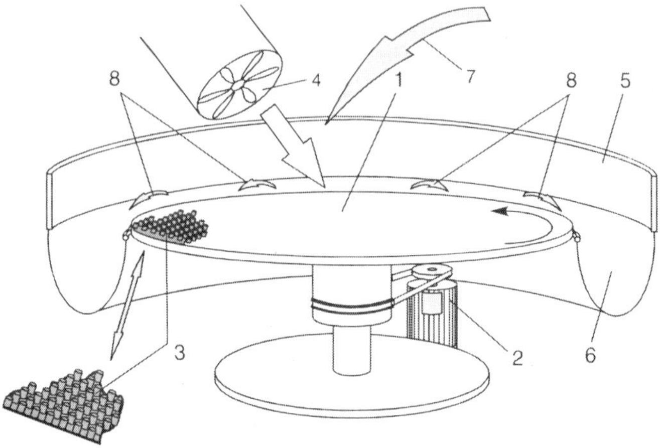

FIGURE 4.10. Device for separation of stigmas and stamens from saffron flowers. 1–round rotating plane, 2–electromotor–belt drive, 3–rubber-nipple cover, 4–ventilator, 5–wall, 6–textile hopper for petals, 7–flowers feeding, 8–flow of petals.

flower due to weak bonds. Higher airflow resistance of petals causes the airflow to bring them to the textile hopper-channel (6) formed around the rotating table (1). The stigmas and stamens fall down and get caught between rubber nipples (3). From time to time, the operator stops the ventilator and drives (2), takes the cover (3) and shakes out the collected stigmas and stamens.

POSTDRYING PROCESSES

Postdrying processing includes diverse operations aimed at providing the final product, material for packaging, or refined raw material for additional processing/finalizing. In most cases, the first operation, if not done before drying, is removal of plant parts—threshing. This is, for example, the removal of stalks and stems from leaves. Size reduction is a frequent operation in many production lines. Sizing sometimes provides the final material for some teas or spices. Separation and classification is a common operation primarily aimed at removing the undesirable parts of plant material, but also for classification of desirable parts according to certain characteristics. There are also some other operations commonly practiced, for example, mixing, pressing, and the like.

Plant Parts Removal—Threshing

The desirable part of the plant depends primarily on the content and composition of the active ingredients. There should be no admixture of other plant parts that do not contain active ingredients, or in which their concentration is very low; moreover, the content of other materials is defined in adequate national and international documentation—pharmacopoeias and standards. With the exception of the food safety legislation, which should be strictly followed, other quality characteristics are changeable according to market demands and fashion. In many cases, especially if the commodity is not a market-ready product, the quality demands could be rather different from the standard demands. But, in almost all production procedures, the removal of undesirable plant parts is expected. This procedure is typical for herbaceous plants as is the removal of stalks and stems from leaves. Sometimes the term leaves removal is used, although the leaves are in almost all cases usable plant parts. Other plant parts are seldom removed, for example, bark, blossoms parts, and the like. The removal of undesirable plant part residuals, for example, stems, belongs to this operation too. Any removal of plant parts is done by mechanical treatment and needs a force to release the bonds between desirable and undesirable parts. Removal is mostly based on rubbing, which implies friction effects. Application of rubbing is generally connected with the following problems:

1. If the strength of the bonding element, for example, stem, is higher than the strength of the bonded parts themselves, the threshing causes their destruction at the same time. Sometimes the destruction of the main parts, for example, leaves, is even more intensive than that of the bonding element.
2. Friction, as a generator of threshing force, destroys active ingredient holders and implicates quality losses.
3. The chopping–disintegration of plant material can also cause difficulties for prompt or later separation. For example, the same level of size reduction of stalks and leaves complicates separation by sieving.

4. Retaining of some parts of the leaves on stalks and stems directly increases losses.
5. Mechanical treatment can generate the changing of color and other positive organoleptic characteristics of desirable plant material.

All this complicates identification of the best solution for plant material threshing, especially if a wide palette of plants has to be processed. Based on the problems listed, the following list of general demands for successful threshing has been established:

1. Full detachment between desirable and undesirable parts of plant material, that is, disconnection of bonds.
2. Limiting the losses of active ingredients.
3. Limiting the degradation of organoleptic characteristics of plant material.
4. Reduction to negligible amount of losses of desirable material still bonded to the undesirable parts.
5. Limiting the share of undesirable material bonded to desirable; limiting the share of impurities.
6. Limiting the share of undersized parts of desirable material, not usable for market products due to improper sizes.
7. Differentiation between particle size of desirable and undesirable material, as a prerequisite for efficient separation by sieving.
8. Avoiding interlaced mixture of desirable and undesirable material, prerequisite for successful separation.
9. Good particle distribution of desirable plant material, for adequate market products.

The threshing will be more successful if more of the listed demands can be fulfilled. It is difficult to establish a uniform evaluation procedure due to other impacts. For example, the producer processes several plants at the same time, and the threshing facility performs differently with different plants. Some of them are not positive enough, but best results are achieved for the most important plant; the facility can fulfill most of the demands, for example, all except demand 3—organoleptic characteristics are degraded. If this characteristic is not important for a buyer, this demand can be neglected. It can be concluded that the threshing should be performed to fulfill diverse product quality and processing demands. From the engineering point of view, the detachment of bonds should be realized with minimal force and with minimal rubbing generated during the process.

Manual threshing is still practiced in many developing countries, especially in small enterprises. There are a few other procedures, but the most common one is the moving of a bunch of material by hand over a wire net/sieve. The worker can adjust the needed pressure exerted between material and net, according to the leaves removal progress. Most often leaves and some parts of broken stalks/stems fall through the net.

Threshing process has been investigated for a long period, primarily for cereals. Although many researchers tried to make mathematical model of threshing, they were unsuccessful. Positive research accomplishments are mostly based on the results of numerous tests. Systematization of threshing tools made for cereals can be used for MAP. Figure 4.11 shows two basic types of threshing devices: tangential and axial.

Threshers presented in Figure 4.11 have high efficiency, but in both cases the rubbing of material is very intensive. Rubbing not only causes size reduction of material but also intensive destruction of active ingredient holders when processing of some materials.

Numerous variations of the basic threshers are possible. Figure 4.12 presents two commonly used threshers with some specific details.

One example of the tangential thresher type shown in Figure 4.11a is also schematically presented in Figure 4.13. This threshing device

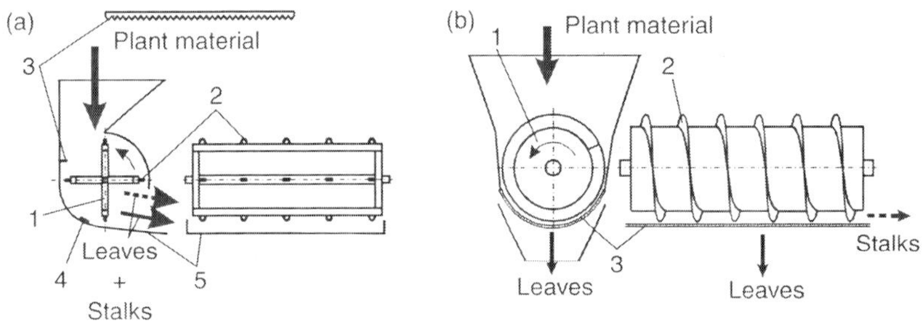

FIGURE 4.11. Commonly used threshing devices, for cereals and MAP. (a) Tangential type: 1–drum-rotor, 2–laths, example, 3–teeth type counter lath, 4–flat type counter lath, 5–bottom housing, (b) Axial type: 1–cylinder, 2–auger, 3–sieve bottom.

is simple and effective. The threshing is generated by the impacts of laths. One, two, or more counter laths intensify the impact effects. The number and type of laths and drum RPM can be changed to achieve adequate parameters for different plants. For the sieve, five diverse screen types could be used. The use of groepel *(frog mouth)* sieves is typical, and it is commonly used in combine harvesters. This sieve separates threshed leaves from stalks/stems. Under the sieve, a vibrating collecting box (6) is located. Figure 4.14 shows the thresh-

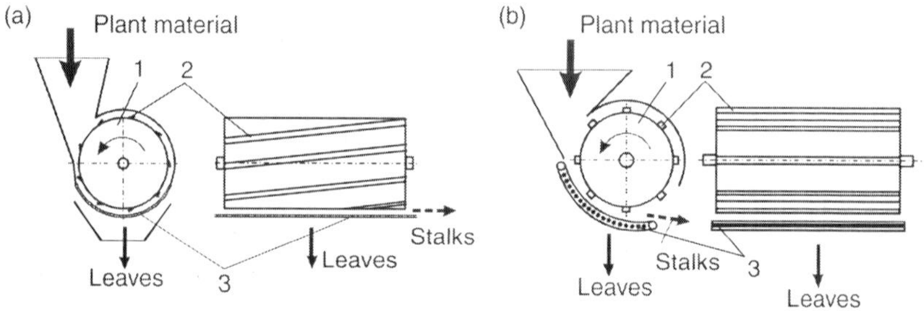

FIGURE 4.12. Examples of commonly used axial and tangential threshers. (a) axial thresher with spiral laths, 1–drum, 2–spiral laths, 3–sieve, (b) Tangential thresher with concave, 1–drum, 2–laths, 3–concave.

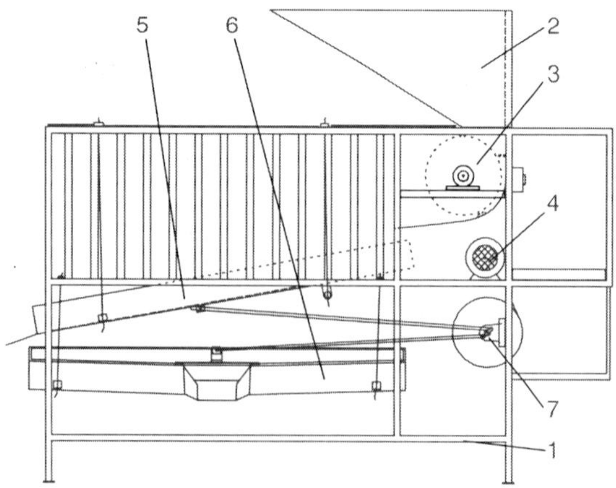

FIGURE 4.13. Thresher with tangential type of threshing device. 1–frame, 2–inlet hopper, 3–threshing device, 4–drive, 5–sieve, 6–collecting box, 7–two eccentrics.

FIGURE 4.14. Threshing effect of the tangential type of threshing device by processing of peppermint and *Origanum dubium*.

ing effects of this machine by processing peppermint and *Origanum dubium*.

Due to the controlled intensity of the strokes and the absence of points of intensive rubbing, the material damages are lower than when using other threshing types. The size of the leaves and stalks is not intensively reduced. Due to minimum of unnecessary dragging, compressing, and rubbing of material, this machine gives positive results concerning defined demands.

There are several threshers with an axial type of threshing device. The scheme of one thresher based on the device shown in Figure 4.11b is shown in Figure 4.15. The drum-rotor (3), is made in the form of a sectional auger. The space between the auger inner wedge and central tube has an important role. If the material is too intensively loaded, or a batch created, it can go through this sector and so avoid overloading. For the threshing of some plants an auger transport is effective enough, but for some others additional stress is needed. Therefore, spiral laths or spiral brushes could be used, Figure 4.16.

The scheme of spiral laths and sieves during the threshing process is shown in Figure 4.17. Depending on the opening dimension and

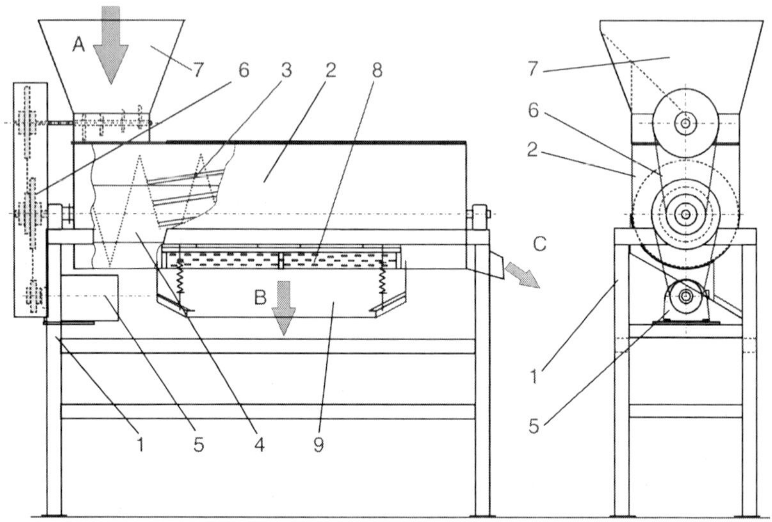

FIGURE 4.15. Axial type thresher device. A–material inlet, B–leaves outlet, C–stalks-stems outlet; 1–frame, 2–tub, 3–drum-rotor, 4–central tube, 5–drive, 6–transmission, 7–inlet hopper, 8–sieve, 9–leaves slope.

form, more or less intensive threshing will be performed. Simultaneously the leaves and stalks will be crumbled. The crumbling intensity depends on various parameters. The same machine can be used as a leaves grinder.

In any case, the sieve has the role of a concave or counter body. Depending on plant properties, different type of sieves and openings can be used. For some plants, stretched metal is adequate, with positive effects to product quality. If the bonds between leaves and stalks are too strong, or plants have some specific architecture, the material should be processed in more than one pass. Depending on the plant type and sieve used, the grinding/crumbling of leaves can be realized with the same machine. The example of material threshed using this machine is presented in Figure 4.18. Some test results of threshing and crumbling of hyssop, thyme, and peppermint are given in Table 4.1.

The stalks and leaves can be separated by sieving or by airflow. The losses of leaves, share of leaves in stalks (outlet C, Figure 4.15), can be avoided by multipassing. The lowest losses have been achieved by

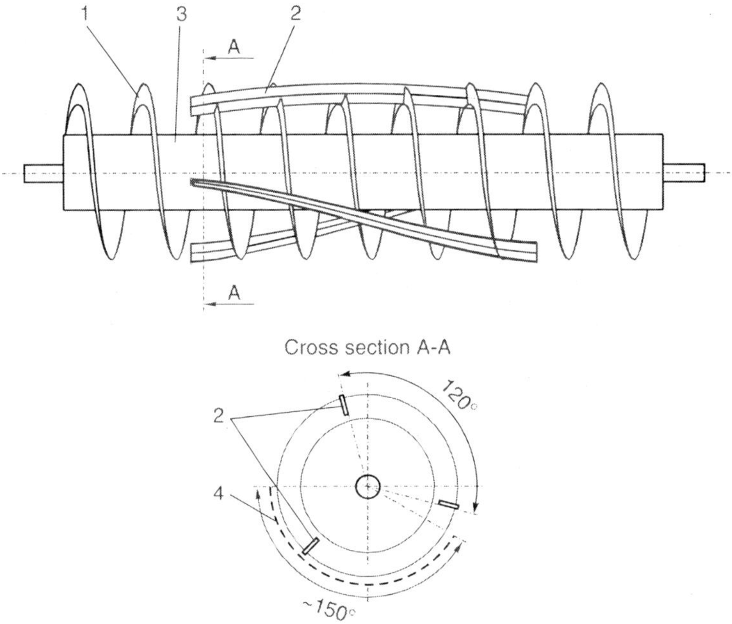

FIGURE 4.16. Drum with axial moving of material and auger details. 1–zonal auger, 2–laths or brushes, 3–central tube, 4–enveloping sieve.

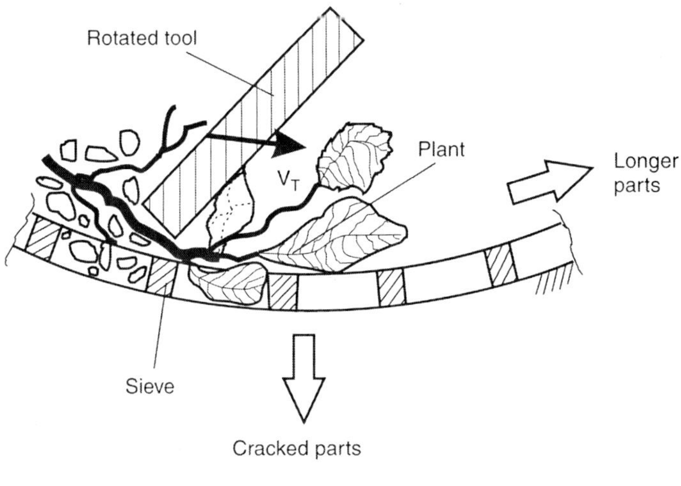

FIGURE 4.17. Spiral lath function.

FIGURE 4.18. Example of auger type threshing: threshed thyme stalks.

TABLE 4.1. Results of peppermint, thyme, and hyssop threshing using the axial type thresher shown in Figure 4.15.

		Thyme		
	Peppermint	Flowering phase	Vegetative phase	Hyssop
Share of stalks, %				
1	18.1	31.2	41.7	24.3
2	40.4	43.8	53.2	33.0
3	26.3	45.1	57.1	37.0
Share of leaves, %				
1	8.5	12.6	4.1	7.4
2	3.6	5.4	1.6	3.8
3	4.2	6.1	2.3	4.2

Source: Veselinov, 2003.

using sieve 3 made of stretched metal. Two-pass processing of peppermint and thyme in the vegetative phase was necessary to reduce losses below 1 percent, while three passes were enough for hyssop and thyme in the flowering phase.

It is very important to use dried material, with equilibrium moisture content. If wet material is treated, the threshing effects will be partially or totally unsatisfactory. Hereinafter, a particular thresher used in practice will be described. The first example is a special type of thresher, an original idea of one innovative producer, shown in Figure 4.19. Here the material is moved over a round stretch of metal surface, pushed by two paddles. The diameter of the working space is 2 m. Leaves and crushed stems/stalks fall down through the stretch metal grate (2) and slide to the lower outlet (12). Stalks with residual leaves leave the machine through the upper outlet (11).

The material, for example, peppermint, has to be treated twice until complete removal of leaves is achieved. The machine can be driven by electromotor or by tractor PTO. The advantage of the machine is its very high capacity, up to 2 t of dry material per hour. It enables quick threshing of the batch dryer content. The time needed for drying of leaves is shorter than the time needed for other plant parts. It is highly recommended to end the process of drying after this point is reached. The leaves have to be removed within approximately 1h.

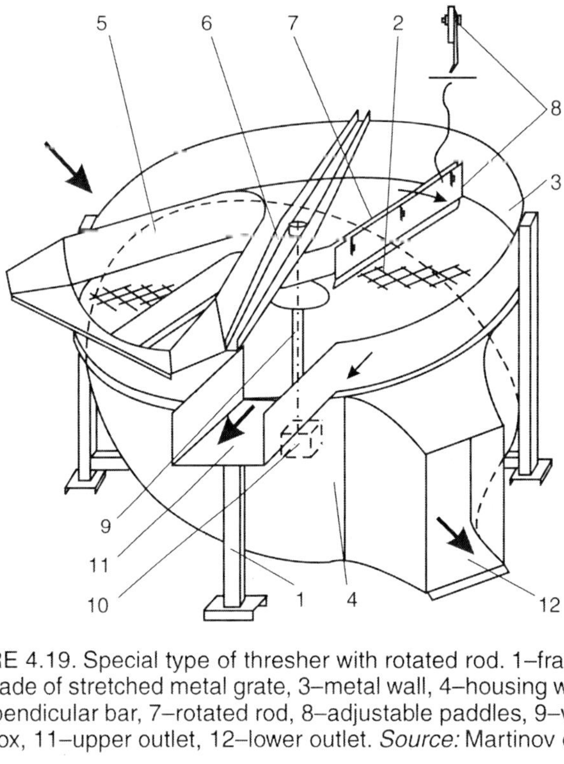

FIGURE 4.19. Special type of thresher with rotated rod. 1–frame, 2–active surface made of stretched metal grate, 3–metal wall, 4–housing with slope, 5–inlet, 6–perpendicular bar, 7–rotated rod, 8–adjustable paddles, 9–vertical shaft, 10–gear box, 11–upper outlet, 12–lower outlet. *Source:* Martinov et al., 1991.

After this period the leaves are again moist, due to moisture migration from other plant parts. The shortening of the drying time, that is, ending the process at the point where the leaves are dry, contributes to the higher capacity of the dryer and reduces fuel consumption. Some essential data for the testing of this machine are given in Table 4.2.

Two passes are performed for mechanical threshing. It is not possible to notice any significant difference between mechanical and manual threshing. Manual threshing results in higher share of whole leaves, but also in higher share of losses of stalks and leaves.

For seed processing, different types of fine threshers are used, primarily for the removal of chaff. Some of them can be used for the threshing of certain MAP, or for additional removing of stems. Examples are given in Figure 4.20 and Figure 4.21. Figure 4.20 shows a tangential type of threshing device, where the drum is formed of brushes and a concave wire net. Many manufacturers of seed processing equipment offer this type of machine. It can be successfully used for plants with weaker bonds between leaves and stalks, for additional removal of stems, or even for leaves grinding. The rubbing of the plant material is realized between brushes (7) and the enveloping exchangeable wire net (6). The openings of the wire net define threshing effects.

In the machine shown in Figure 4.21, a cone form of wire net is used. In addition, axial movement of the shaft (4) can adjust the distance between the brushes and the net to reduce the rubbing of

TABLE 4.2. Results of testing thresher with rotated rod compared with manual threshing.

	Share in % of[a]				
Threshing procedure	**Whole leaves**	**Cracked leaves**	**Leaves with stalks**	**Stalks**	**Leaves losses in %[b]**
Thresher RPM					
100	1.7	63.3	14.2	10.8	3.8
130	15.7	71.1	7.7	5.5	3.5
Manual	22.8	42.3	13.2	21.7	5.4

Source: Martinov et al., 1991.
[a]lower outlet-leaves
[b]upper outlet-stalks

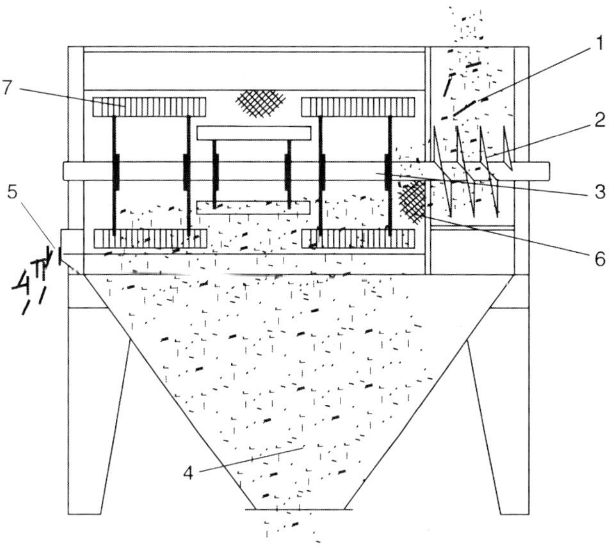

FIGURE 4.20. Thresher with brushes and cylindrical wire net. 1–inlet, 2–feeding auger, 3–shaft, 4–outlet for smaller particles, 5–outlet for bigger particles, 6–exchangeable wires net, 7–brushes.

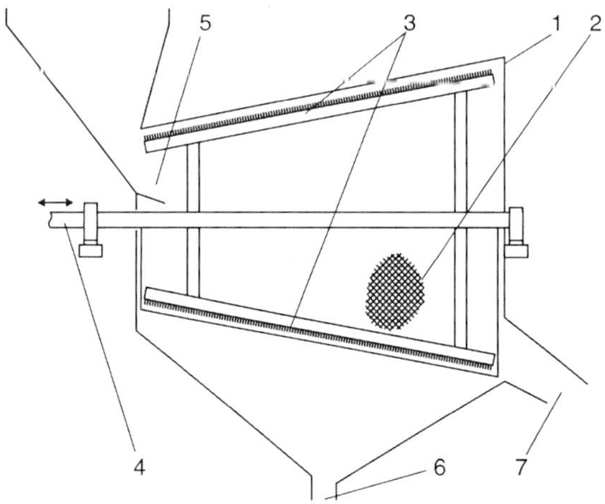

FIGURE 4.21. Threshing machine with brushes and cone wire net. 1–housing, 2–half cone wire net, 3–cone formed brushes, 4–axially adjustable shaft, 5–inlet, 6–outlet for fine particles, 7–outlet for bigger particles.

smaller particles. This threshing machine can be used not only for the processing of threshed material to remove the stems, but also for other cleaning operations.

Particle Size Reduction

The term "size reduction" implies all processes aimed at reducing the size of plant material—primarily to get the size suitable for market products, e.g., tea, or as one additional processing procedure. There are generally three different procedures for mechanical size reduction:

1. Cutting
2. Crushing/crumbling
3. Milling

Cutting

Cutting is dominantly a process of material shearing, Figure 4.22. The material is here treated with moving edge–blade tool (1). The material is supported by stable plane ground (2). In praxis, a counter-blade is frequently used as a support. The material structure is destroyed by shear stress, which is the dominant stress in this process. Figure 4.22a presents the perpendicular motion of the cutting knife, while Figure 4.22b presents the inclined one. Cutting with an inclined knife enables a more effective and "cleaner" cut due to the slipping effect, that is, velocity component v_t. The stresses in the plant material are complex. The first phase of cutting of the material is compressing; the knife blade moves down, causing pressing stress and layer height reduction. When the pressing stress exceeds the material shear strength, the cutting begins.

Most cutting machines have a counter-blade, Figure 4.23. The cutting is realized between two blades that act like scissors. Due to the phase of material pressing, the needed cutting energy is decreased (Tesic, 1984). Implications of material pressing by knife are generation of material layer expansion with angle β_E and dispersion of particles of cut material. Prepressing of forage materials results in lower

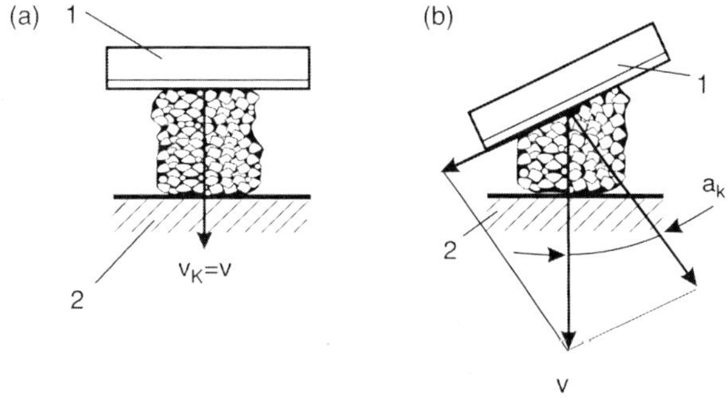

FIGURE 4.22. Plant material cutting. (a) perpendicular cut, 1–knife, 2–ground, v_K–knife velocity; (b) inclined cut.

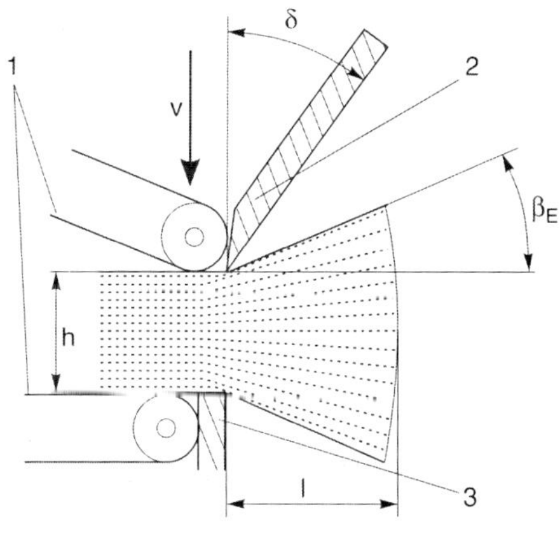

FIGURE 4.23. Typical cutting device for plant materials. 1–prepressing device, 2–knife, 3–counter, δ–cut angle, v–cut velocity, β_E–material layer expansion angle, h–height of compressed material layer, l–cutting length.

cutting energy. In the case of MAP cutting the pressing should be avoided to a higher extent due to its negative influence on active ingredients. This has a consequence of higher energy input that is here less important than the quality.

There are two types of cutters: free-cut and exact-cut devices. With the free-cut devices the length of the material coming into contact with the knife is neither adjusted nor controlled. Exact-cut devices have auxiliaries for adjustable and controlled feeding of material. It enables adjustment of the cutting length of the plant material layer.

Crushing/Crumbling

A schematic description of the crushing/crumbling process is given in Figure 4.24. The material is here placed between stable and movable tool parts, whereby the tools have no edges and pressure stresses are primarily generated. The velocity of the moving part is v_M, and the impact force is F. When the strength of the material is exceeded and brittle reduction cracks develop, the material is destroyed into smaller parts. For some materials with high brittleness, crushing can be realized only by pressing between two bodies, for example, two rollers. It is obvious that the crumbling is possible only if the material is not tenacious but brittle. Most of the dry MAP comply with this prerequisite.

The crushing process is neither predictable nor controllable. Effects are strongly dependent on device parameters and plant material properties. Every MAP has to be pretested to select the most suitable processing operation, i.e., device. If a certain size reduction has to be achieved, in most cases a sieve has to be used or multipassing practiced. An important characteristic of crumbling is that there is high impact of cracking and low impact of rubbing.

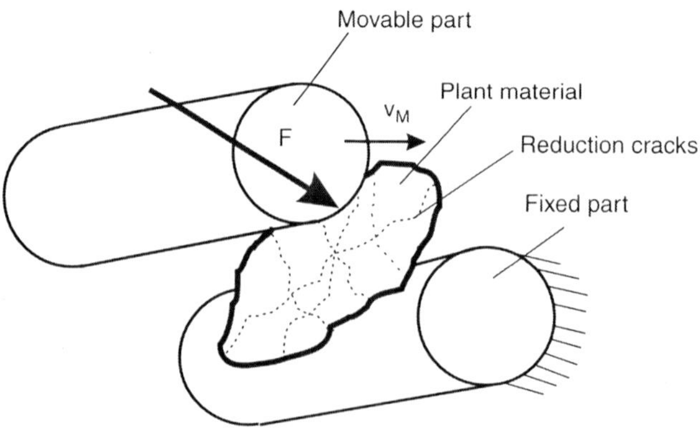

FIGURE 4.24. Material crushing/crumbling.

Milling

The principle of particle size reduction by milling is primarily the rubbing of the material between two abrasive bodies. This is mostly practiced to obtain flour or some powder material from grains. Another milling device, much more used in MAP processing, is the hammer mill. A genuine hammer mill consists of flywheel hammer bodies that rotate with high RPMs and process material by smashing. The only resistant force in smashing is the inertia of the material. In most of the facilities in use, there are some other stationary auxiliaries used for the generation of resistance force, e.g., sieve in Figure 4.25.

Hammer impact, together with the sieve resistance forces, causes material crumbling. The negative effect is that the material is moved between the sieve and the hammer, if not immediately smashed. That causes intensive rubbing that could be avoided by some special measures, for example, larger distance between hammers and sieve. This relatively simple machine can be efficiently used for MAP production, but only for dry, brittle, crumbly materials.

Cutting Devices

Exact-cut cutting devices are well known and used for forage harvesting. Figure 4.26 shows the two most common types. The cutting

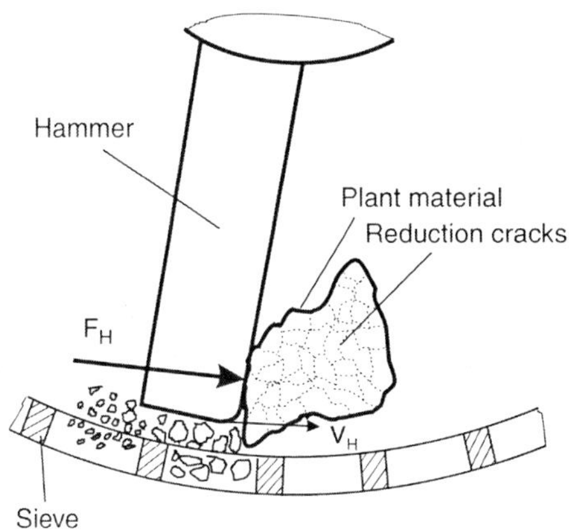

FIGURE 4.25. Particle cracking in hammer mill with sieve.

length is easily adjusted by the combination of feeding velocity and knife frequency or drum RPM. Actually, the problem is material orientation by feeding. An extreme example would represent the stalks or other longer plant parts being fed perpendicularly which makes cutting impossible. Improvement of the feeding device is necessary for both types. Instead of aggressive metal bands, soft rubber should be used. The intensity of pressing of the upper band should be much lower and easily adjustable, as it is usually done for forages processing.

Flywheel type, Figure 4.26a, has spiral knives for MAP processing to achieve a smooth and gentle cut of sensitive plant material. This type of cutting machine has been adjusted for MAP; some elements and assemblies were redesigned and used for predrying cutting of fresh material, Figure 4.26b. The drum type, Figure 4.26b, is commonly used for forage harvesting and only occasionally for MAP cutting, for example, before distillation/extraction.

The special type of exact-cut with parallel moving knives, primarily used for tobacco processing and redesigned for MAP, is schemati-

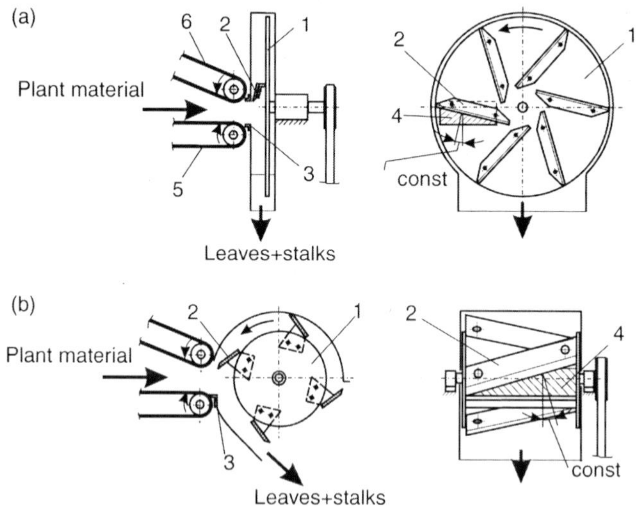

FIGURE 4.26. Two types of exact-cut devices commonly used for forages. (a) Flywheel type: 1–fly wheel, 2–knife, 3–counter knife, 4–sharpener, 5–feeding band, 6–press band, (b) Drum type: 1–drum, 2–knives, 3–counter knife, 4–material layer cross section.

cally presented in Figure 4.27. A vertically moving knife has a special trajectory. It slides down very close to the counter-knife and moves up much further from it. In that manner, an unhampered feeding of material is enabled. The second peculiarity is inclined perpendicular knives (2) that move simultaneously with the "main" knife. These knives are placed under the main knife. The perpendicular knives are made of metal bars. The distance between knives is adjustable to obtain the needed size of particles. Knives (2) make a perpendicular cut along the material layer, and, in combination with the longitudinal cut of the "main" knife (1) result in a rectangular (or quadrate–square, as a special case) form of cut material. Because of that in Germany, and other countries, this machine is sometimes called a "quadrate cut machine."

The feeding mechanism is also adapted according to the special requirements of MAP processing. Both bands are made of soft rubber. The position and spring load of the upper band are adjustable. Both bands are moved with intermittence. During the process of cutting the bands are stationary and do not disturb the cutting process. The material is fed when the knife unit moves up. This type of machine is shown in Figure 4.28. In the newest version of this machine, the upper band is supported by five rollers, without a spring load.

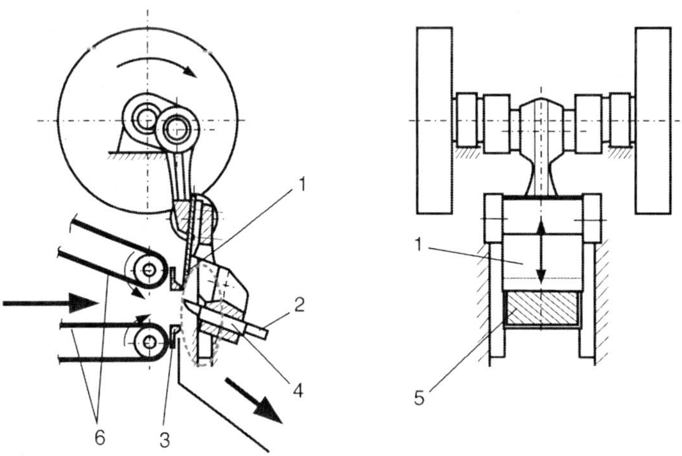

FIGURE 4.27. Special cutter with parallelly moved knives. 1–knife, 2–perpendicular knives, 3–counter-knife, 4–distance adjustment parts, 5–material layer cross section, 6–upper and lower feeding band.

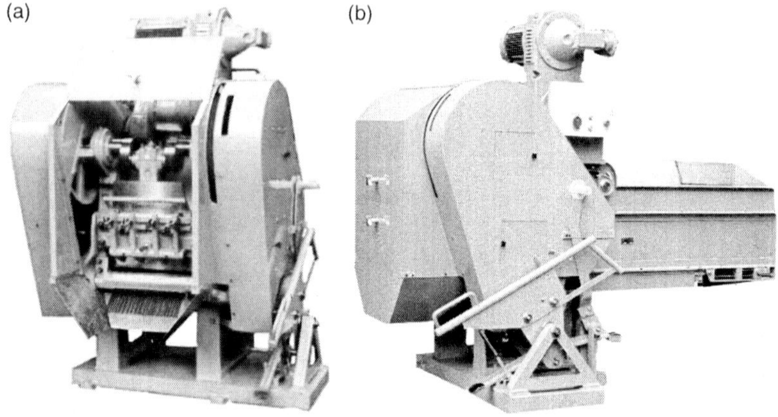

FIGURE 4.28. Market-available "rectangular cutting" machine.

The cutting quality of this type of machine is high, especially regarding the low essential oil reduction and adequate particle dimensions. The demerit of this machine is the high price due to the high engineering and manufacturing demands. This cutter is commonly used in medium and large-scale companies in developed countries.

Two typical free-cut types of cutters are shown in Figure 4.29. The cutter shown in Figure 4.29a is typically used for the cutting of garden residues. The length of the cutting and particle size are completely uncontrolled and depend only on the feeding rate. This simple machine can be successfully used in material preparation for extraction or distillation. The cutting device shown in Figure 4.29b enables indirect control of particle size using sieve (4). The sieve prevents the fall-through of material larger than its openings.

The cutter presented in Figure 4.30 shows a simpler version of the cutting machine shown in Figure 4.29b. The rotor (4) with three knives and two counter-knives (3) placed on both sides is used. Under the rotor with knives, an exchangeable sieve (5) is mounted at a certain distance. The sieve openings define the resulting particle size. This cutter is used very efficiently, not only for some grains grinding, but also for leaves.

Detailed cross section of the free-cut cutter type in Figure 4.31 gives more details about the machine shown in Figure 4.29. This type

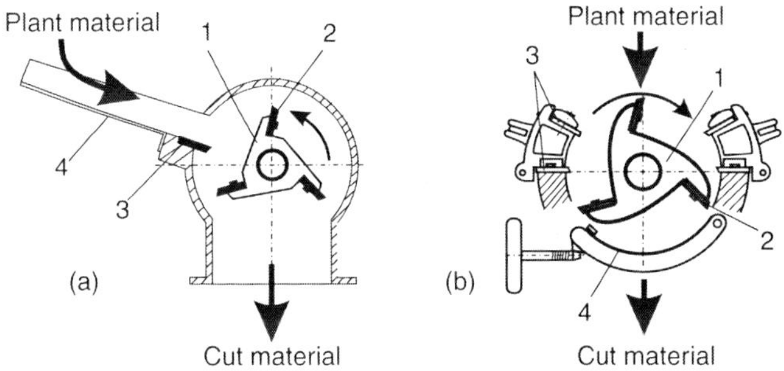

FIGURE 4.29. Free-cut types of cutters. (a) with free fall of material: 1–rotor, 2–knives, 3–counter knife, 4–feeding platform, (b) with sieve: 1–rotor, 2–knives, 3–counter knives, 4–sieve.

is also widely used for other materials, for example, plastic granules. Here are shown two assemblies of counter-knives, left (3) and right (4). The adjustment of the clearance between the knives and counter-knives needs the special attention of the operators. To achieve the right gap, commonly from 0.1 mm to 0.15 mm, all knives are longitudinally movable. First, the gap between one of the counter-knives and the rotor knife with the biggest radius is adjusted. After that, the other two rotors knives are adjusted according to the first counter-knife. Finally, the other three counter knives are to be adjusted according to the rotor knives. These adjustments have to be done precisely to ensure efficient cutting. This makes the machine complex and expensive.

Different type of sieves (6), perforated sheet metal or wire net, with different openings is fixed to a swinging frame (5). The distance between the knife edges and the sieve is adjustable to enable the flow of different material amounts, but it is also limited to a certain minimum. Here the undesirable mechanical treatment of plant material is reduced to the minimum, because no prepressing is applied. The use of finely adjusted knife pairs results in a smooth cut. With the proper choice of sieve type, openings, and distance between knife edges and the sieve, controlled size reduction can be achieved. One example of a machine equipped with this type of cutting device is shown in Figure 4.32.

Some other types of free-cut devices are also available for MAP processing, but it should be always be pretested and evaluated according to defined demands. There are many special types of MAP

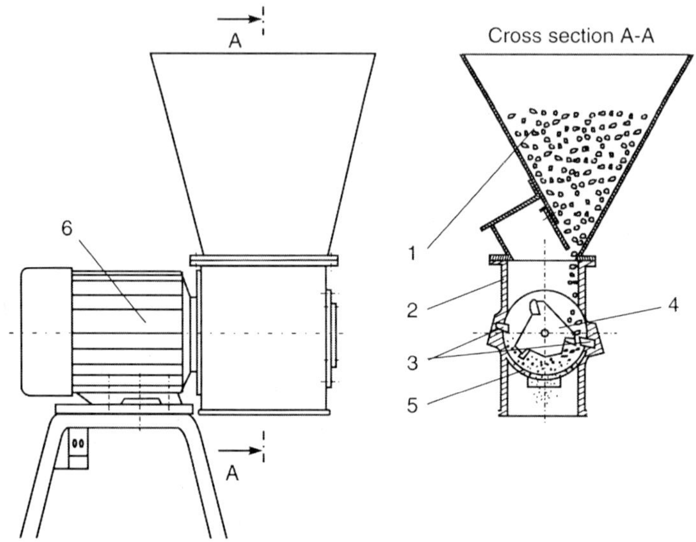

FIGURE 4.30. Free-cut cutter type–simpler solution. 1–hopper, 2–housing, 3–counter-knives, 4–rotor with knives, 5–sieve, 6–drive.

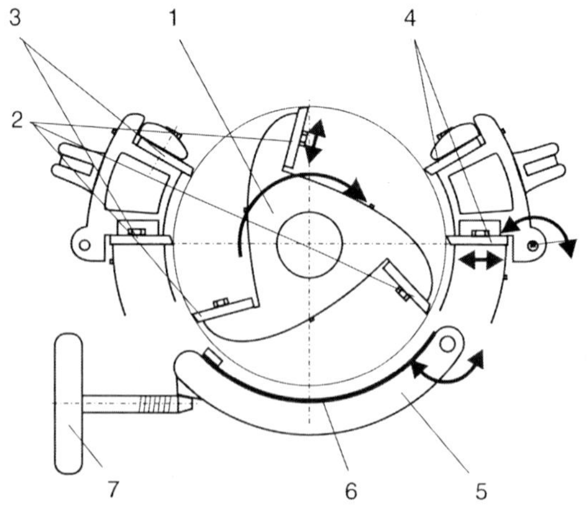

FIGURE 4.31. Free-cut cutting device with sieve. 1–rotor, 2–knives, 3–left side counter-knives, 4–right side counter knives, 5–swinging frame, 6–exchangeable sieve, 7–sieve positioning.

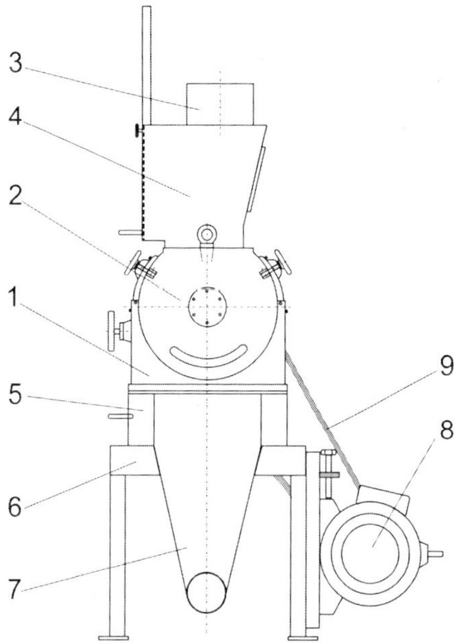

FIGURE 4.32. Machine with free-cut cutting device with sieve. 1–housing, 2–chopping unit, 3–feeding tube, 4–hopper, 5–outlet box, 6–frame, 7–duct, 8–electromotor, 9–belt drive.

cutters developed by MAP or farm machinery producers. An example of a two-cylinders cutting device is shown in Figure 4.33.

Here the edges of the grooves have the function of the counter-knife. The material comes from above and only gravity generates the pressure force. This enables gentle treatment of plant material. The cutter can be used for fresh and dry material, Figure 4.33a. For different cutting lengths, an appropriate pair of cylinders should be used, Figure 4.33b and c. For the cutting of larger pieces, supporting fingers are mounted on a smooth cylinder, Figure 4.33c. The use of different pairs of cylinders for different cutting lengths causes higher costs and repetitive reassembling. A similar type of cutter has been developed and tested for the cutting of marshmallow bars, schematically presented in Figure 4.34. Instead of the counter-cylinder, a plastic panel (3) is used here. The cutting disc knives (2) penetrate some 1 to 1.5 mm into the hard plastic panel, forming grooves. The cutting

(a) (b) (c)

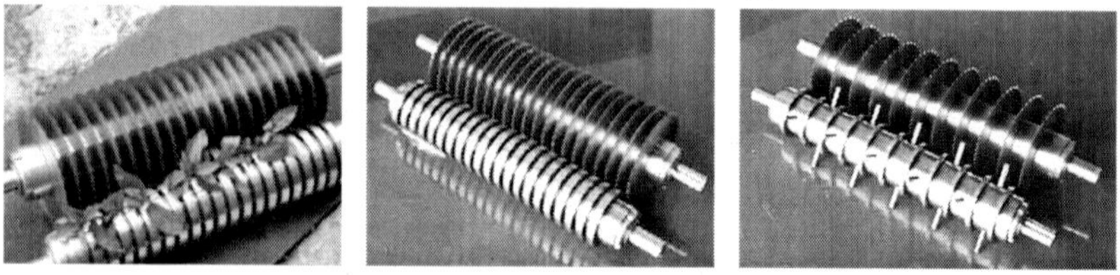

FIGURE 4.33. Special two-cylinder cutting device for MAP. (a) Function of cutter: 1–cylinder with cutting discs (top), 2–smooth cylinder with perpendicular grooves (bottom), (b) Cylinder set for 20 mm cut length, (c) Cylinder set for 40 mm cut length. *Source:* Anonymous, 2003.

width can be adjusted by the selection of different distance rings (4). The cutting results are good, especially if fresh material is used.

A special type of cutter aimed at shortening or removal of flower stems is schematically presented in Figure 4.35. The cell drums (2) are designed to enable enough space for flowers, detailed in Figure 4.35. Knives (1) are mounted on an oscillating rod (5) that is constantly tensed downwards by springs (3) to enable tight position of knives and edges of cell drums. The oscillating mechanism moves the oscillating rod with knives.

The machine was tested and it showed very good results for stem cutting of chamomile, but it can be used for other flowers, as well. One example of this type of machine, with integrated special type of sieve beneath the cell drums, is shown in Figure 4.36.

Cutters are usually too complicated to be produced in small workshops. Some elements and assemblies have to be produced with high accuracy and special steel type have to be used.

Crushing Machines

There are different types of crushers used for medicinal and aromatic plants. Most machines show similar results as threshers and in some cases could be used for both operations, for example, the axial type thresher device shown in Figure 4.15. The machine can be treated both as thresher and crushing machine depending on the more significant effect for certain plants or procedures. A simple crushing/grinding unit for very brittle material is schematically presented in Figure 4.37. The pressing of material between cylinder (1) and band

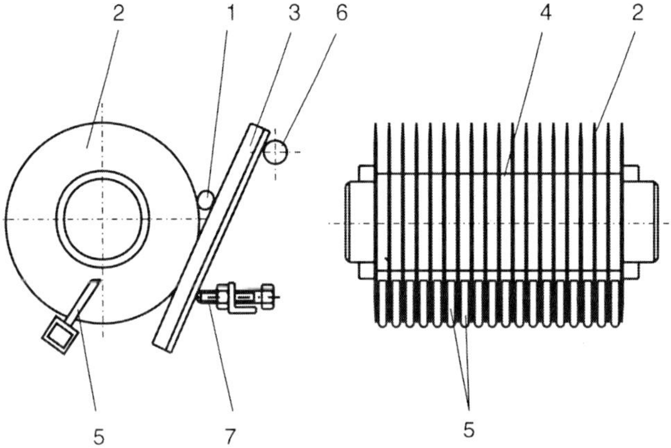

FIGURE 4.34. Cutter for marshmallow bars. 1–marshmallow bar, 2–cutting discs, 3–inclined plastic panel, 4–distance rings, 5–scrapers-cleaners, 6–stable hinge, 7–positioning screw. *Source:* Sijacic, 1999.

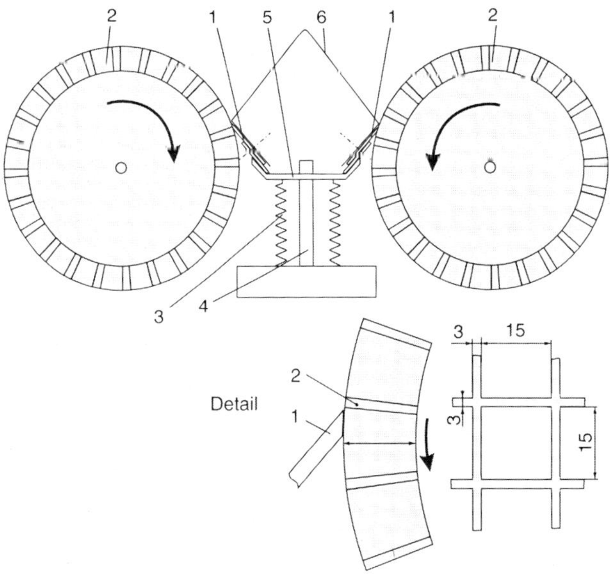

FIGURE 4.35. Flower stem cutter. 1–knife, 2–cell drum, 3–tension springs, 4–liner, 5–oscillating rod, 6–divider.

FIGURE 4.36. Flower stem cutter.

conveyor (2) causes crushing and size reduction, primarily of leaves, because they are more brittle in general. The intensity of size reduction can be controlled by the selection of cylinder type, for example, the corrugation pattern, cylinder band distance, and band/cylinder velocity. This device could also be used for stems removal if they are not too brittle.

An example of a crusher used for size reduction of MAP in small and medium-size enterprises is given in Figure 4.38. This is an axial type crushing device, similar to the axial thresher with spiral laths shown in Figure 4.12a. When compared with the thresher, the clearance between the drum and drum housing is smaller.

A sieve is integrated into the machine. Replacement of sieves under the drum is usually possible, to enable using the machine for different plant materials and different cutting sizes. The axial type crushing machine and an example of its crushing results are presented in Figure 4.39. There are three different particle size groups. The finest one can be used for bag teas, and the other two for diverse products. In some cases, multipassing is also practiced.

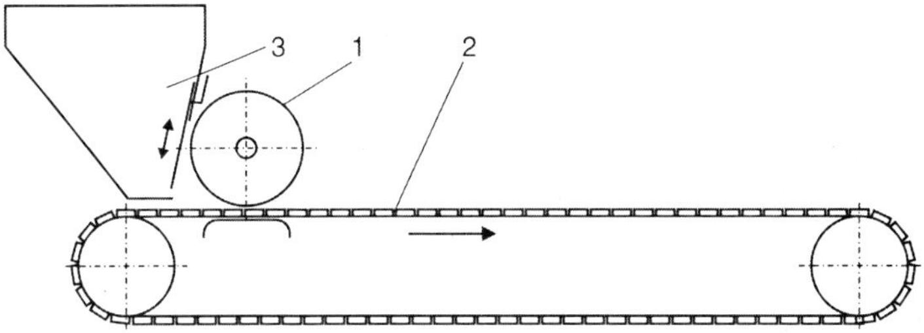

FIGURE 4.37. Simple MAP crushing unit. 1–smooth or corrugated cylinders, 2–transversally supported band conveyor, 3–hopper with dozing slider.

Milling Facilities

Due to their simple design, low purchase price, and a more universal usage for other crops, hammer mills are often used for MAP. The most common types of hammer mills are shown in Figure 4.40. The first one, Figure 4.40a, is not a typical hammer mill, because a set of fixed beaters has been used. This is nowadays a very popular solution, for example, a simple and inexpensive facility for hay processing. Blunt or sharp beaters or their combination could be used, as shown in Figure 4.40a. The counter laths additionally mill plant material from outside the sieve. The dimension of the sieve opening determines the particle size. If the material is not sensitive, that is, mechanical processing will not destroy the holders of the active ingredients, for example, grain materials, the use of the hammer mill can be very efficient and favorable if very small particles are demanded. The use of this type of machine for the processing of sensitive plants is very limited. Hammer mills can be successfully used for material preparation for extraction, distillation, and similar operations, if performed immediately after milling. In other cases, milling can cause extremely high essential oil losses. The process of milling is also energy-consumptive. There are some special hammer mill designs available. One of the special mills for MAP is shown in Figure 4.41. Here, flying thin hammers made of stainless steel are used. The impact of these hammers causes favorable size reduction of plant material. The exchangeable sieves are placed at a certain distance from the hammers and therefore the rubbing effect is reduced.

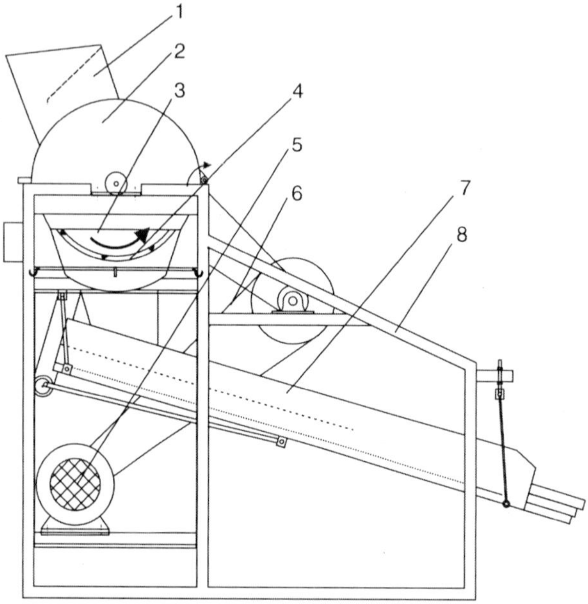

FIGURE 4.38. Axial type crushing machine used by small and medium-size companies. 1–hopper, 2–drum housing, 3–drum with spiral laths, 4–sieve, 5–electromotor, 6–belt transmission, 7–sieve, 8–frame.

Special Facilities for Size Reduction

Some of the machines used for size reduction of MAP materials cannot be classified into any of the defined and listed groups. The examples given here are hybrids of two or more of them: dog rose crusher, Figure 4.42, and combined grinder mill, Figure 4.43.

The machine in Figure 4.42 is aimed at crushing the body of dog rose to enable successful separation of fruits, pericarp and needles, whereby additional size reduction is not desirable. High RPM rotated beaters-knives (4) crush axially fed dog roses. The main generator of crushing is the direct impact, but the counter knife (6), also contributes. The sieve (5) is intended to stop bigger particles from falling down. The bigger particles are kept inside the mill for additional processing. This facility is a combination of cutter and hammer mill with fixed hammers.

FIGURE 4.39. Axial type crushing machine and the results of peppermint processing.

The machine presented in Figure 4.41 is commonly used for the grinding of spices, although it could be used for some medicinal plants too. It is remarkably good for grains grinding. The machine processes the material by a combination of crumbling and milling. The beaters are fastened to the hub (1). The front side or mill cover (4) and the opposite side to the cover (2) are made of profiled plates with radial grooves. The sieve (3) has a minor influence on the material processing but enables passing of particles smaller than the opening diameter.

The principles of grinding in this machine are based on very intensive rubbing. It means that this machine could be successfully used only for rubbing insensitive plant material, for example, grains.

Postdrying Separation and Classification

Separation and classification is a common operation in almost every food production chain. Removal of all, or the most undesirable,

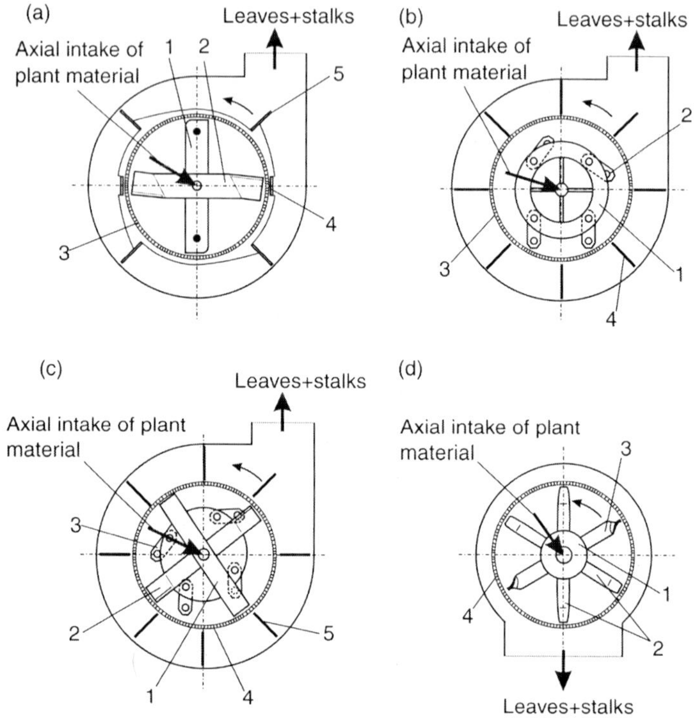

FIGURE 4.40. Hammer mills–the most common types. (a) With fixed beaters and pneumatic transport: 1–blunt beater, 2–blade beater, 3–sieve, 4–counter laths, 5–impeller, (b) With flying beaters: 1–rotor, 2–flying beaters, 3–sieve, 4–impeller, (c) With combined beaters: 1–blunt beater, 2–knife beater, 3–flying beaters, 4–sieve, 5–impeller, (d) With fixed beaters and gravity transport–falling: 1–rotor, 2–type 1 knives, 3–type 2 knives, 4–sieve.

materials and plant parts, as well as sorting according to some characteristic, is a very important refining process that was accomplished manually in the past. Manual refining processes are mostly abandoned now and are practiced today only in certain regions and for certain plants. At the same time, manual inspection is still a common procedure even in highly mechanized facilities, where inspection bands or tables are used. Mechanized separation and classification are always based on the difference in certain characteristics of components within the material. The prerequisite for the separation of

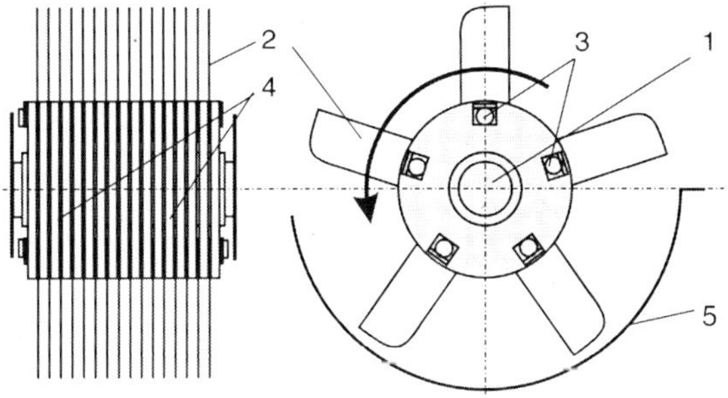

FIGURE 4.41. Example of specially designed hammer mill for MAP. 1–shaft, 2–stainless steel hammers, 3–screws, 4–distant plates, 5–sieve.

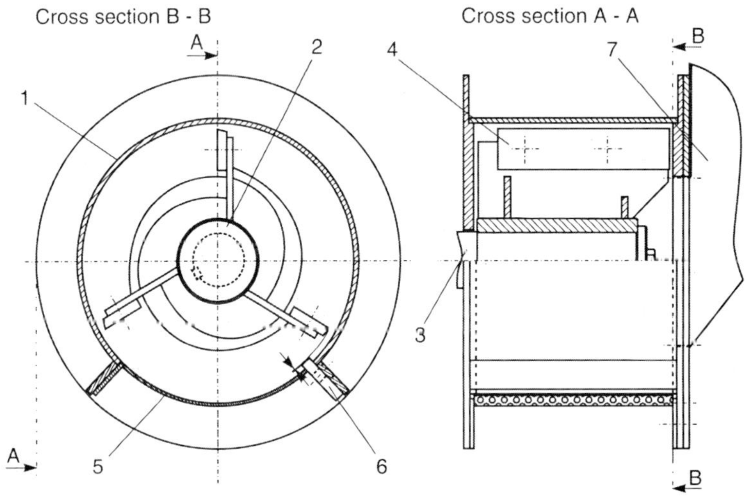

FIGURE 4.42. Dog rose (Rosa Canina) crusher. 1–housing, 2–rotor, 3–shaft, 4–beaters-knives, 5–sieve, 6–counter knife, 7–hopper. *Source:* Konstantinovic, 2000.

any undesirable/desirable component is that some characteristic of that component differs enough to enable its separation from the rest of the material. Figure 4.44 illustrates this demand.

Before the separation-classification procedure is chosen, one should know the fundamental data on material components. If some of the

FIGURE 4.43. Combined grinder-mill for MAP—machine view with open door. 1–hub with beaters, 2–radial grooves backside, 3–sieve, 4–radial grooves on the cover.

characteristics of the undesirable components differ, as shown in Figure 4.44a, successful separation is possible. The result of the separation, of course, depends on the quality of the device and procedure used. If the fields of characteristics of components that have to be separated are partly overlapping, Figure 4.44b, complete separation is not possible, even if the best device and procedure are used. In this case, it is necessary to make a compromise by choosing the facility either to have losses of desirable material or to have an admixture of undesirable components. If the overlapping of fields of characteristics is significant, the separation based on this material characteristic difference is not possible, Figure 4.44c. For example, if the critical velocity ranges from 3.3 to 4.2 ms^{-1} for leaves, that is, from 7.8 to 8.7 ms^{-1}, for stalks, then successful airflow separation is possible; if the critical velocity ranges from 7.5 to 8.9 ms^{-1}, for grains, that is, from 8.1 to 9.2 ms^{-1}, for admixed particles, airflow separation is not possible.

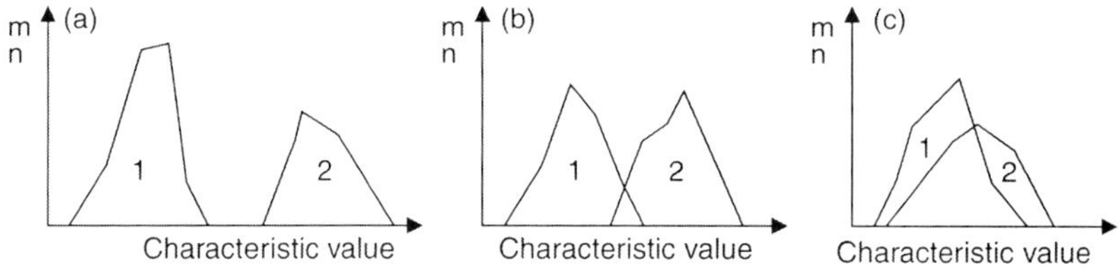

FIGURE 4.44. Characteristics of material components and prerequisite for separation, mass distribution (m) or number distribution (n). (a) fields of characteristics that are distinctly apart, (b) the fields of characteristics that are partly overlapping, (c) the fields of characteristics that are significantly overlapping.

Diverse physical properties could be used for this operation. Some of them are given here, listed according to the importance and frequency of usage:

1. Size, dimensions
2. Air flow resistance
3. Form, length—cross section relation
4. Coefficient of friction
5. Rolling resistance
6. Density
7. Electric properties
8. Color

Separation-Classification According to Size/Dimensions

There are many plant material mixtures consisting of components with different sizes/dimensions. It enables separation of undesirable or desirable parts using facilities that are able to recognize this difference. The same procedure can be used for material classification. A special procedure is also removal of undersized or oversized components in desirable material.

The most common devices for separation-classification using differences in size/dimension are different types of sieves. The material moves over sieve surface and smaller particles fall downwards through openings, while the bigger ones continue moving. The sieve opening determines this process. Due to the particle shape and sieve—particle dynamic process of sieving—it is not always unique and unambiguous. Figure 4.45 shows the moving and falling of particles through the sieve opening by one mode of sieve motion.

Sieving effect strongly depends on the form of particles. The simple definition: *opening should be slightly bigger than the diameter* is valid only for slow-moving spherical-form particles. For example, for stalk-like form, that is, where one dimension is much bigger than the other two, the particles will fall down if the center of gravity and one end of the stalk are in the opening.

For small ratio r/l, the mathematical description of particle dynamic is:

$$x = r(1 - \cos\omega t) \quad \dot{x} = r\omega \sin\omega t \quad \ddot{x} = r\omega^2 \cos\omega t.$$

Whereas the inertial force of particle is:

$$F_i = -m \cdot r\omega^2 \cos\omega t.$$

For the case in Figure 4.45 (a), there follows:

$$\pm G\sin\alpha + F_i \cos(\varepsilon \pm \alpha) + F_A \cos(\gamma \pm \alpha) > F_t \text{ and}$$
$$F_N = G\cos\alpha - F_i \sin(\varepsilon \pm \alpha) - F_A \sin(\gamma \pm \alpha).$$

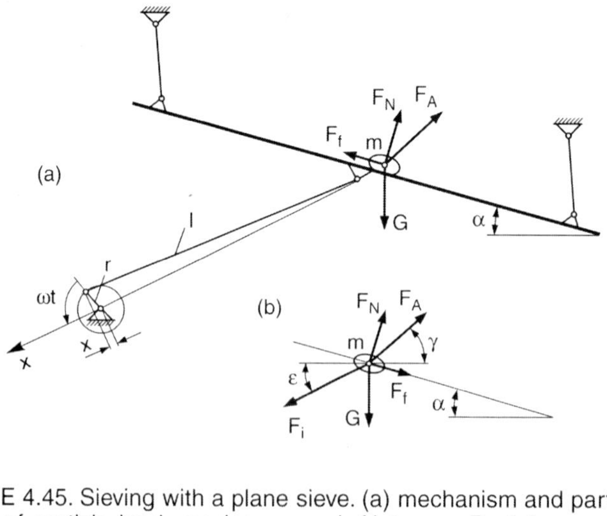

FIGURE 4.45. Sieving with a plane sieve. (a) mechanism and particle load, (b) details of particle load, moving upward. *Notes:* m: Particle mass; G: Particle weight; F_N: Force normal to seive; F_r: Friction force; direction opposive of travel; F_A: Airflow resistance of particle; F_i: Inertia force of particle; φ: Angle of friction particle sieve.

Conditional equation for particle movement from the beginning to the end of the sieve is:

$$F_i \cos(\varphi \mp \alpha - \varepsilon) > G\sin(\varphi \mp \alpha) - F_A \cos(\varphi \mp \alpha - \gamma)$$

$$mrw_1^2 \cos\omega_1 t \cos(\varphi \mp \alpha - \varepsilon) > G\sin(\varphi \mp \alpha) - F_A \cos(\varphi \mp \alpha - \gamma),$$

or for $g = 9.8\ ms^{-1}$ is

$$\frac{rw_1^2}{9.8}\cos\omega_1 t = k_1 \cos\omega_1 t > \frac{\sin(\varphi \mp \alpha)}{\cos(\varphi \mp \alpha - \varepsilon)} - \frac{F_A \cos(\varphi \mp \alpha - \gamma)}{G\cos(\varphi \mp \alpha - \varepsilon)};$$

here is $k_1 = \dfrac{r\omega_1^2}{9.8}$.

For sieving without airflow, $F_A = 0$. Coefficient k_1 is:

$$k_1 > \frac{\sin(\varphi \pm \alpha)}{\cos(\varphi \pm \alpha - \varepsilon)\cos\omega_1 t}.$$

For particles moving upward and downward the coefficient k_2 for the same conditions will be:

$$k_2 > \frac{\sin(\varphi \pm \alpha)}{\cos(\varphi \pm \alpha - \varepsilon)\cos\omega_2 t}.$$

Or for the jumping of particles, the coefficient k_3 is:

$$k_3 > \frac{\cos\varphi}{\sin(\varepsilon \pm \alpha)\cos\omega_3 t}.$$

Grain sieving requires:

$$k > k_3 > k_2 > k_1.$$

This theoretical approach is not easy to be realized and is not applicable for some new sieve designs. A simplified description of sieve oscillation intensity can also be:

$$k = \frac{ew^2}{g}$$

where k is oscillation intensity factor, e is half amplitude, ω is frequency, and g is gravity acceleration. For MAP processing, the intensity factor k of 1.06 to 1.10 is sufficient. This does not cause jumping of plant material. A typical example is: $e = 7$ mm, $\omega = 39\ s^{-1}$, $k = 1.085$.

The definition that all particles with maximum dimension smaller than the opening must drop through is very simplified. Figure 4.46 shows such conditions.

Conditions for particle dropping through sieve opening are:

$$x > L, \quad y \geq h \text{ whereby}$$

$$L = \left(S - \frac{1}{2}\right)\cos\alpha - \frac{b}{2}\sin\alpha - \delta\cos\varepsilon.$$

Here δ is sieve relocation during the time of particle dropping t in seconds. Limiting velocity v_L can be calculated as:

$$L \geq v_L \cos\alpha \quad h \leq \frac{gt^2}{2} + v_L t\sin\alpha \quad \delta = 2e$$

$$v_L t\cos\alpha \leq \left(S + \frac{1}{2}\right)\cos\alpha - \frac{b}{2}\sin\alpha - \delta\cos\varepsilon$$

$$\text{if} \quad \text{use } A = S - \frac{1}{2} - \frac{b}{2}\text{tg}\alpha - 2e\frac{\cos\varepsilon}{\cos\alpha}.$$

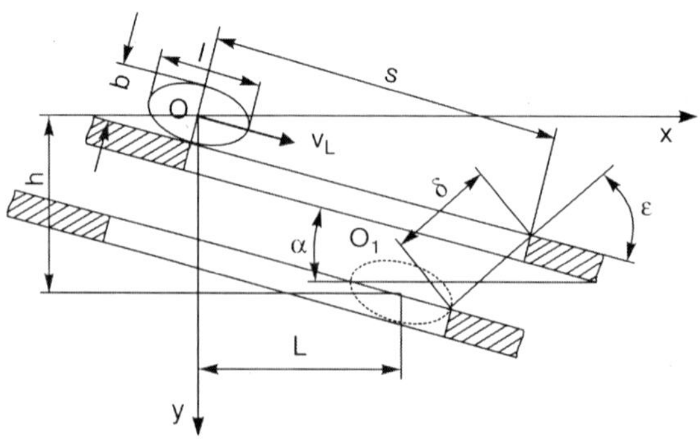

FIGURE 4.46. Particles moving and dropping through sieve opening, limiting velocity.

The limiting velocity value would be:

$$v_L = A\sqrt{\frac{g\cos\alpha}{b + \delta\sin(\varepsilon + \alpha)}}$$

in ms^{-1} if all other SI units were used.

This complicated approach is used here only to illustrate the importance of the selection of e and ω relation to achieve the same value of oscillating intensity factor k. The limiting velocity for grain sieving is in the range of 0.38 to 0.45 ms^{-1}.

The particle velocity also depends on the sieve slope and directly influences the capacity of the sieve. A very simple equation for calculating capacity is:

$$Q = v_{layer} \cdot A_{layer} \cdot$$

Cross section A_{layer} is a product of the width and height. Greater height means more obstacles for smaller particles to drop through the material layer. It is an additional barrier for successful separation-classification. The chances for successful separation-classification are higher if the sieves are longer. That is why the sieves commonly have their length twice the size of the width.

There are two additional criteria for sieve evaluation:

1. Coefficient of separation efficiency ε_E.
2. Coefficient of separation–classification thoroughness ε_C.

The first one is:

$$\varepsilon_E = \frac{m_S}{m_{AP}}$$

where m_S is the mass of separated undesirable/desirable particles and m_{AP} is the total mass of undesirable/desirable particles. This can be estimated by testing 3 to 5 samples of sieved material using a laboratory sieve. If the value is greater than 0.8, the separation is classified as successful, while values below 0.5 qualify as unsuccessful.

The second coefficient is:

$$\varepsilon_C = \frac{m_C}{m_T}$$

where m_C is the mass of particles found in certain dimension class, and m_T total mass of this dimension class. The laboratory sieve can be used for this evaluation as well.

Although there are numerous guidelines for sieve selection, successful sieve selection is mostly a matter of experience, primarily due to the very different behavior of different plants and even plant parts.

For R&D activities, and for some commercial purposes, determining particle size can be very important. For this purpose different methods are available. Some of them are standardized, for example, by the standards given by ASAE (Anonymous, 1993). Laboratory sieves are used for these purposes. Larger laboratory sieves are commonly perforated, and the smaller ones are made of net wire. If there is a need to determine the content and distribution of very small particles, for example, under 0.1 mm, computer analysis of microscopic digital photos could be successfully practiced.

Particle distribution is mostly expressed through cumulative mass for defined size groups. If it is not differently defined, normal distribution or log-normal distribution paper is used. That enables description of particle distribution by only two parameters, median *(m)* and standard deviation *(s)*. It enables additional elaboration of size distribution characteristics, e.g., the expected share of particles is smaller or bigger than a certain size value. Figure 4.47 shows examples of two particle size distributions on log-normal paper, with the same value of the median.

In Figure 4.47 line A presents distribution with lower standard deviation, that is, smaller range of particle size. For line B the range is wider, which causes higher content of particles smaller than particle size value, for example, for 1.3 mm about 8 percent for B and 2.6 percent for A, and for 25 mm about 99.1 percent for B and 94 percent for A.

There are certain problems associated with the correct analysis of sieving results. For example, particle shape influences dropping of material particles through sieve openings. The problem is to define

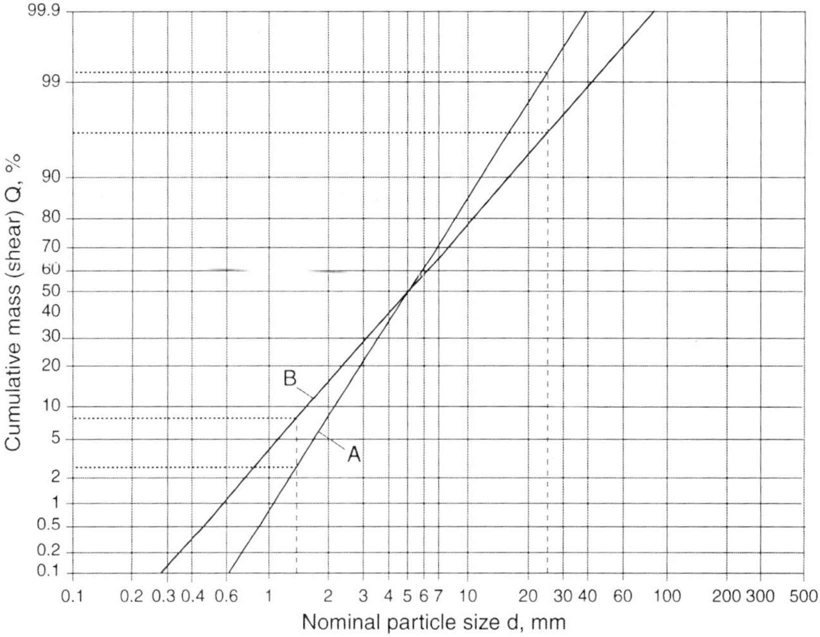

FIGURE 4.47. Presentation of particle size distribution on log-normal paper.

the representative size of particles found in one size group, for example, between sieves with openings of 5 and 8 mm. As shown in Figure 4.46, the particles having length almost twice that of the opening diameter, can drop through, especially if their diameter is negligible in comparison with length, for example, stalks. If the absolute value of the particle size is not important, but the results are used to compare the size distribution of two materials, the analysis is satisfying. If not, the representative values of the size group have to be corrected. To draw a line through points representing test data is also a considerable problem, which can be overcome by using regression analysis, whereby the same method can be used for the evaluation of the distribution pattern (Veselinov, 2003).

Sieves could be divided in two groups: plane and cylinder. The plane sieves are more widely used and the following description is related primarily to them. Some details concerning cylinder sieve type will also follow. The most significant characteristic of a sieve is its opening size. If the openings are not round or square (dominated shapes), some additional data will be needed. The thickness of the sieve is also an influencing parameter. The forms of the sieves are different, but most are rectangular and round, Figure 4.48. The forms of the openings are mostly round, perforated in sheet metal, or square, made of wire net. For very small openings, under 0.5 mm, plastic woven net is used. The different shapes of the sieve openings are perforated in metal and used for the separation of longer grains and stems. Most sieves are fastened to the frame by welding, soldering, or splicing. Only thicker sieves are sometimes used without an additional frame or some other supporting structure.

New technical achievements enable new possibilities for sieve design, which would provide the appropriate particle motion and obtain better sieving results. Possible sieve oscillating modes are shown in Figure 4.49. The first mode, a typical mode for conventional sieves, is oscillating in a vertical plane whereby the velocity vector could have an angle α within the range $0 < \alpha < 90°$ with the sieve plane. The second, $\alpha = 90°$, and the third, $\alpha = 0°$, are special cases. The fourth mode is when a sieve moves horizontally, left and right. The fifth mode presents the rotating motion of a sieve in a horizontal plane, and the sixth motion of the sieve in a vertical plane.

The oscillating mode in combination with intensity could have a great impact on the processing of some plant material mixtures. For example, if modes 1 or 2 are used, and the oscillating intensity

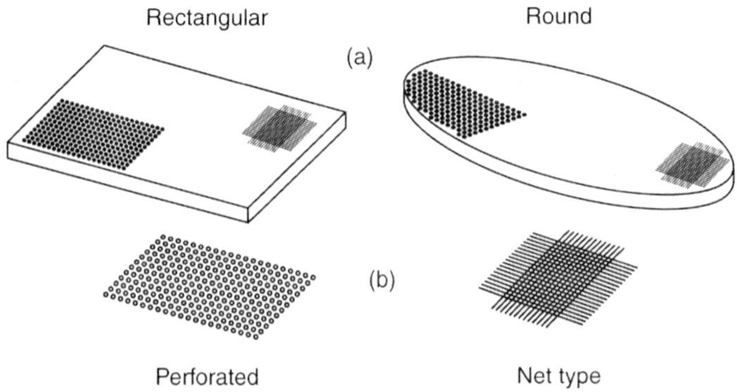

FIGURE 4.48. The most common sieve shape (a) and opening types (b).

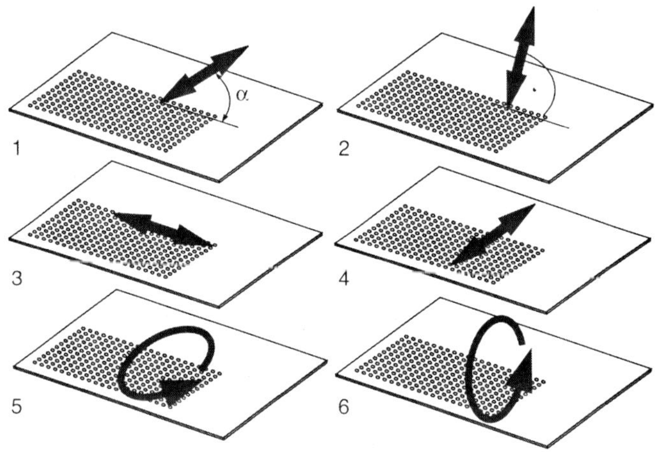

FIGURE 4.49. Oscillating modes of sieve.

generates jumping of material, there is nothing to force the stalks and stems to fall through the sieve openings. In this case, the axles of the stalks and stems are in a vertical position and the cross-section dimensions are smaller than the dimensions of the opening. The sieving of some ellipsoid grains makes this effect positive and it provides a higher capacity of sieving.

The shaking intensity of the sieve is an important factor. Higher intensity can generate intensive material fluidization with flying particle effect. In that case, the particles cannot fall through the openings. Lower intensity shaking enables smaller particles to travel over the sieve on top of bigger ones, without the possibility of falling through the sieve openings. Changing the shaking intensity by regulation of amplitude or oscillating frequency solves this problem. Today this is possible with the use of the frequency converter even for simpler and cheaper sieving devices.

The sieves can be configured as follows:

1. Vertical–one above the other
2. Horizontal–one behind the other
3. Combined

The most widespread are vertically configured sieves. The biggest opening is on the top. Among the horizontally configured sieves, the first sieve has the smallest openings, and the last one has the biggest. That can cause problems, because the smallest particles should find a path through the whole layer of material at the beginning of sieving. The other demerit of horizontally configured sieves is that the sieve is longer—all sieve lengths are summed up. Due to longer frames and housing of the sieve, it is difficult to eliminate negative vibration effects. The merit of this sieve configuration is that it enables inspection of the flow of the material and also allows manual sieve cleaning. Combined sieves are used for special machines and demands.

There are diverse types of oscillation generators with different types of linkage to the frame. The most typical are the following:

1. Vertical eccentric system
2. Horizontal eccentric system
3. Electromotor with unbalanced mass
4. Vertical axis rotated disc with unbalanced mass
5. Horizontal axis rotated disc with unbalanced mass
6. Electromagnetic exciter

For simple machines, eccentric electromotor-driven systems are frequently used. The exception is the system driven by a tractor PTO or other similar engines. In that case, the linkages between sieve housing and frame are less important, and only some designs have the additional task of reducing vibration transfer as a buffer.

Almost all sieves with unbalanced mass have linkages with spring characteristics to enable the generation of oscillations of the sieve housing, whereby diverse materials and designs are used. The adjustment of oscillations intensity is achieved by changing the position of the center of mass and RPM.

Oscillations can have a significant impact on the machine structure and environment. Depending on the intensity of the oscillations, buffering is either necessary or desirable. There are a few methods for buffering:

1. Reduction of sieve and housing mass to the possible minimum
2. Coupling of sieves and housing with counter oscillating mass, for example use of two housings with opposite oscillating directions
3. Use of strong massive frames fastened to the ground

The failure of sieve elements occurs frequently, especially in the simpler models. There are some measures to avoid this, for example, by the use of wooden bars or elastic elements, which support the changing of the moving direction of the housing with sieves. Sometimes the strengthening of a frame results in negative effects. Some new materials with integrated buffering effects can solve this problem without high investments and interventions.

One of the most typical sieves is shown in Figure 4.50. This sieve has been known for a long time and is still in use, especially in developing countries, by small and medium-scale producers. Here, a vertical eccentric mechanism is used to generate the oscillations of the sieve housing.

This type of sieve consists of a vertically configured sieve, which operates in the first mode of oscillation, as can be seen in Figure 4.49. Oscillating arms (4) are usually made of hard wood, and an elastic pin-joint (10) is used to reduce the negative influence of the oscillations to the machine elements. The fastening of the frame to reduce the negative influence of the vibrations can cause a problem. The connection to the ground should not be too strong. The best possible solution would be a good balanced machine, with low vibration intensity and without the need to fix it but only to limit its motion.

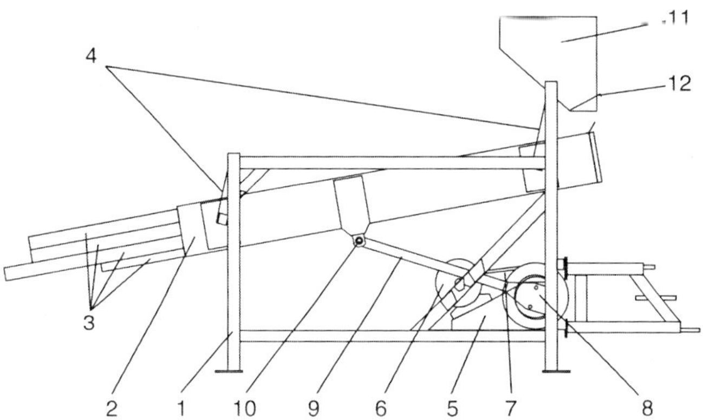

FIGURE 4.50. Sieve with vertical eccentric mechanism. 1–frame, 2–sieve housing, 3–fraction outlets, 4–oscillating arms, 5–drive housing, 6–eccentric disc, 7–belt, 8–drive unit, 9–rod, 10–pin-joint, 11–hopper, 12–slider.

Another example is a sieve with a horizontal eccentric mechanism, Figure 4.51. Here, a vertical shaft is used to rotate a plate with eccentric mass. The coupled plate is connected to the sieve box by a spherical bearing. The plate is fixed to the gravity center of the box, that is, in the center point between two vertical steel cables. The rotation of the discs with steel cables enables changing of the sieve box slope. The change of slope changes the material flow, separation process, and capacity of sieve.

A sieve with a horizontal eccentric mechanism is shown in Figure 4.52. The lower shield covers the eccentric plates. The upper shield covers the belt drive with balance mass meant to reduce the vibrations of the machine. This type of machine consists of a vertically configured sieve, which operates in the fifth mode of oscillation, as can be seen in Figure 4.49. For the oscillation buffering a counter mass is used. Due to the difference of acting planes a torque is generated that can have negative impacts on the construction.

Electromotors with an unbalanced mass for oscillation generation are nowadays more frequently applied, due to mass production of this

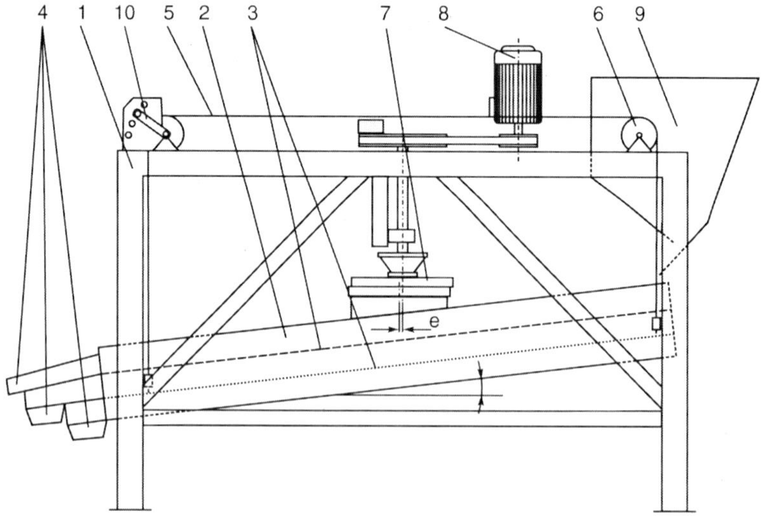

FIGURE 4.51. Sieve with horizontal eccentric mechanism. 1–frame, 2–sieve box, 3–sieves, 4–outlets, 5–steel cables, 6–rollers, 7–eccentric situated plates, 8–drive, 9–hopper with slider, 10–slope adjustment device.

FIGURE 4.52. Sieve with horizontal eccentric mechanism.

unit and the low prices. A low power electromotor has a shaft with both side journals. Two plates are here connected to the adjustable unbalanced mass. A simple sieve using this device for oscillation generation is shown in Figure 4.53.

This type of machine operates in the sixth mode of oscillation, as can be seen in Figure 4.49, or in the fifth mode in some special cases. The intensity of oscillations depends on the characteristics of the rubber elements (5) used for connecting the sieve housing to the frame. For bigger sieves, two oscillation generators are usually used to neutralize side oscillations, Figure 4.54. They are mounted on the sides of the sieve housing. In most cases, the position of the generators can be adjusted by rotation. The position of the generators defines the direction of the oscillations, i.e., pattern of sieve housing motion. If they are more oriented to the sieve plane, the perpendicular oscillations are less intensive, but the material flows higher. A typical use of this motion pattern is for grain cleaning, whereby the airflow is used to separate the chaff and other light impurities. The success of the separation can be observed through the transparent covering of the vertical duct (3).

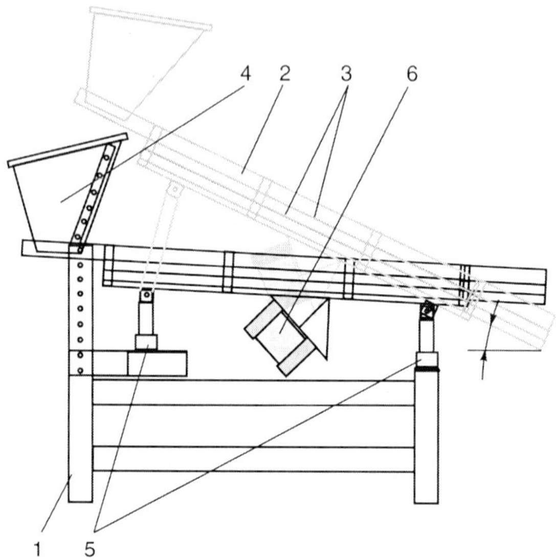

FIGURE 4.53. Sieve with adjustable slope with electromotor and unbalanced mass as oscillation generator. 1–frame, 2–sieve housing, 3–sieves, 4–hopper, 5–rubber link elements, 6–oscillation generator–electromotor with unbalanced mass. *Source:* Anonymous, 2003b. Reprinted with permission from Euro Prima.

The machine with round sieves is schematically presented in Figure 4.55. This type of sieve is used for pulverized material, for example, flour and gypsum. This is actually a mechanized version of a hand sieve used in households, with the oscillation generator, whereby a vertical axis rotated disc with eccentric mass is used. Sieve housing motion in the horizontal plane represents the fifth oscillating mode as can be seen in Figure 4.49.

To use the full potential of the device, different trajectories are applied to move particles over the sieve surface. A special case of primary separation of heavy undersize particles, for example, sand, used for cereals, is presented here. The round sieves are more expensive in comparison with rectangular ones, due to the round form and ledges on the sieve surface needed for material movement control.

The housing of the sieves can be differently connected to the frame. Figure 4.55 shows a connection with springs, or spring material, for example, rubber blocks, and Figure 4.56 shows an example of a connection with steel cables.

FIGURE 4.54. Sieve with two side-mounted generators of oscillations. 1–electromotor with unbalanced mass–oscillations' generators, 2–rubber blocks–connecting sieve housing and frame, 3–vertical duct with transparent covering.

Square sieves and oscillations generator using vertical axis plate with an unbalanced mass that showed very good sieving results is presented in Figure 4.57. This type of sieve operates in the sixth mode of oscillation as can be seen in Figure 4.49. It can be successfully used for the separation and classification of a wide palette of medicinal and aromatic plants. The scheme of a sieve with a horizontal axis rotated disc with unbalanced mass is shown in Figure 4.58 combined with airflow separation of light material parts.

This type of machine operates in the first or the second mode of oscillation as can be seen in Figure 4.49. This type of oscillation is suitable only for grains sieving. It is a typical mistake to purchase this type of sieve as the only device for sieving. The machine is primarily aimed at grain processing where the light parts, chaff and leaves, are trash. The mode of oscillations is not suitable for the separation of

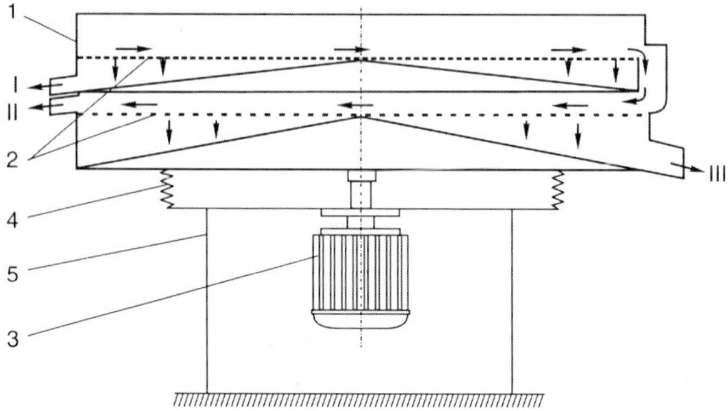

FIGURE 4.55. Sieve with round sieves. 1–sieve housing, 2–sieves, 3–drive, 4–spring links, 5–housing, I–outlet for the smallest parts, II–outlet for usable parts, III–outlet for oversized parts. *Source:* Petrusov, 1975.

stems. The stems could easily jump and perform improper passing through the sieve openings.

An electromagnetic exciter consists of a ferric mass supported by a spring that moves linearly. The ferric mass is attracted by an electromagnet to one side and pulled back by a spring when the electromagnetic field is gone. The forces induced by electromagnetic field are stronger than the forces of the spring. Due to the intermittent electromagnetic field that is switched on and off in certain positions, the ferric mass moves in both directions. The intensity of the oscillations depends on the mass and the path length of the ferric body. This device can generate the first four oscillating modes as can be seen in Figure 4.49.

Some simple sieves, mostly mechanically driven using the vertical eccentric and mode 4 of oscillation are still in use in the developing countries. Some of them have adjustable air flow support. These simple and inexpensive machines can be successfully used for the separation and classification of grainy MAP products, as additional devices.

The drum sieve, using cylindrical sieves, has been described in the section titled "Predrying Separation and Classification" with the example shown in Figure 4.8. There are no principle differences in the sieve used for dry material. Figure 4.59 shows a drum sieve with four cylindrical sieve sections.

FIGURE 4.56. Version of round sieves with steel cable connection of sieve housing and frame.

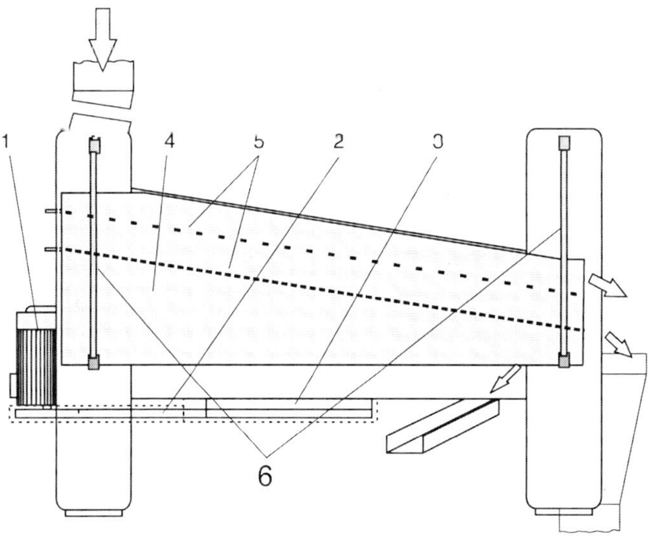

FIGURE 4.57. Sieve using vertical axis rotated disc with unbalanced mass. 1–electromotor, 2–belt transmission, 3–disc with unbalanced mass, 4–sieve housing, 5–sieves, 6–rods connecting sieve housing to the frame.

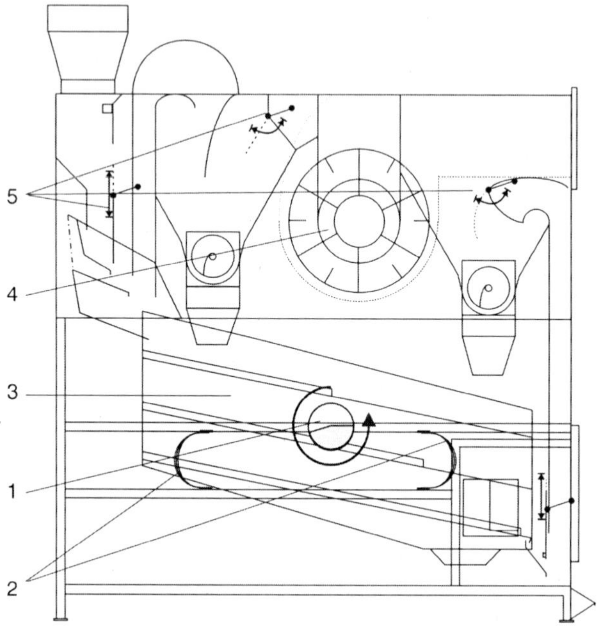

FIGURE 4.58. Sieve with horizontal axis rotated disc with unbalanced mass. 1–two sided mounted discs with unbalanced mass on horizontal driven axes, 2–spring connections between sieve housing and frame, 3–sieve housing, 4–ventilator generated under pressure inside of machine housing, 5–devices for regulation of intensity and direction of air flow.

Due to the gentle removal of material and the possibility of material flow adjustment, the drum sieve could be successfully used for many MAP. The disadvantage of this sieve type is the high price of cylindrical sieves and the problems that emerge when sieves with small openings, less than 2 mm, are used. For a diameter of approximately one meter, the sieve length is three meters. The changing and fastening of sieves is more complicated than with plane sieves. That is why the drum sieves are primarily made for processing of just one type of material, for example, wood chips. Another demerit of the sectional drum sieve type is that the first sieve is always the smallest one. That can cause problems, because the smallest particles should find a path through the whole layer of material at the beginning of the sieving. It can cause slowing down of the sieving process and some errors. The possibility of sieve cleaning is good. Almost sufficient

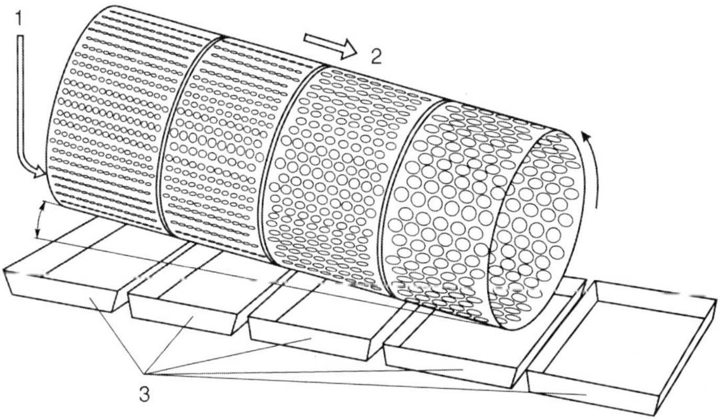

FIGURE 4.59. Drum sieve with four cylindrical sections. 1–material inlet, 2–material flow direction, 3–hoppers. *Source:* Reprinted with permission from Kaltschmitt and Hartmann, 2001.

self-cleaning is possible if longitudinal brushes are applied. An example of a small-scale drum type sieve designed for laboratory testing, supplied with drum diameter of about 0.33 m, is presented in Figure 4.60. In Figure 4.61 the sieve for separation and classification of dry saffron is presented.

The sieve for dry saffron is designed to enable easy exchange of sieves and to be used for testing of sieving effects for diverse material mixtures. That enables preselection of the appropriate sieve openings for sieving applied in the processing facility. This manually driven sieve is primarily designed for the separation of saffron stigmas and stamens.

The cleaning of the sieve is a typical problem. Small inexpensive sieves have no sieve cleaner mounted. The cleaning is practiced in the following manner: the processing is stopped for a while; the sieves are taken out and the cleaning is done using hand brushes or other tools. For some plants or plant parts that cause frequent jamming of the openings, continuous processing is almost impossible. If only the upper sieve jamming is intensive, it can be overcome by frequent manual cleaning. The mechanical cleaning of sieves is a common element of almost all "better" sieves. Rubber balls that jump in a box situated under the sieve are commonly used. The balls pound

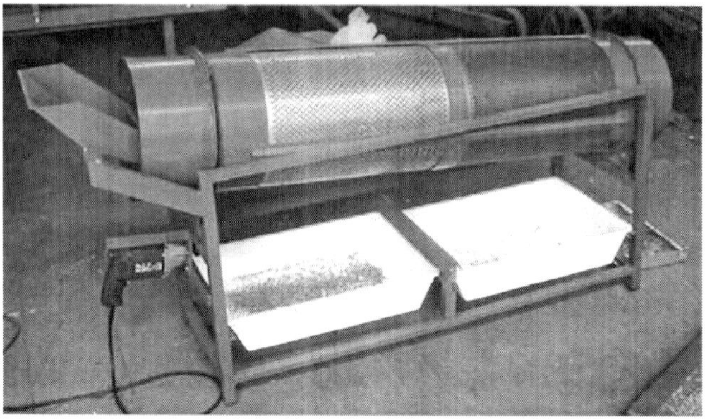

FIGURE 4.60. Small-scale drum type sieve for laboratory usage.

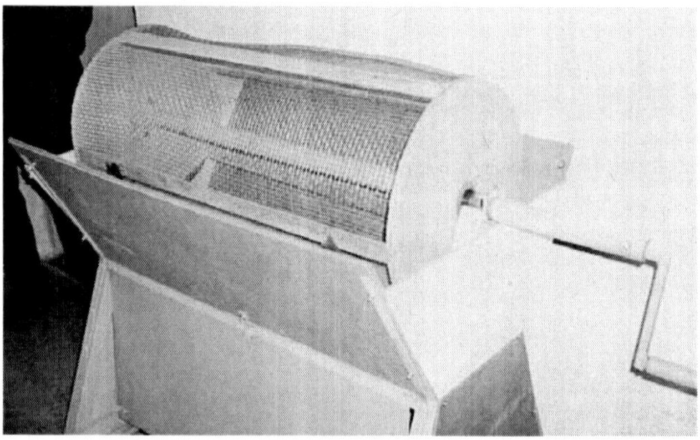

FIGURE 4.61. Drum sieve for dry saffron processing.

the sieves and help remove jammed material from the openings, Figure 4.62.

Another widely used cleaning principle is presented in Figure 4.63. Brushes are mounted on the side belts. The belts are driven and the brushes clean the sieve openings constantly. The moving direction can accelerate or slow down the material flow depending on the process needs, but it can also cause undesirable crumbling of pro-

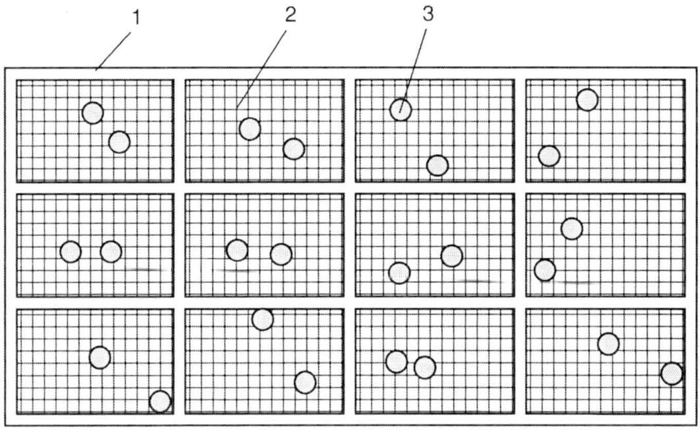

FIGURE 4.62. Sieve cleaning with rubber balls. 1–wooden frame, 2–wire net with big openings, 3–rubber balls.

cessed material. Both types of sieve cleaners are efficient and the price is the only issue to be considered before purchasing a sieving device equipped with one of them.

Airflow Separation and Classification

The difference in the aerodynamic characteristics of plant materials and admixtures is frequently used for separation and classification. The major characteristic is called drag coefficient. Major terms related to the drag coefficient are terminal velocity and critical velocity (Mohsenin, 1980; Grochowicz, 1980). Terminal velocity is fluid velocity by which an object or particle falls down in the fluid flow with constant speed, i.e., with no acceleration or deceleration. The definition of critical velocity is practically the same. That is, the velocity of the fluid oriented upward causes no movement of an object or particle. That means, in both cases, the flow resistance and object weight have the same value. This can be expressed by the equation:

$$G_O = F_R = C \cdot A_p \frac{\rho_f \cdot v_c^2}{2} \text{ or } v_c = \sqrt{\frac{2 \cdot F_R}{C \cdot A_p \cdot \rho_f}} \text{ i.e. } v_c = \sqrt{\frac{2 \cdot G_O}{C \cdot A_p \cdot \rho_f}}$$

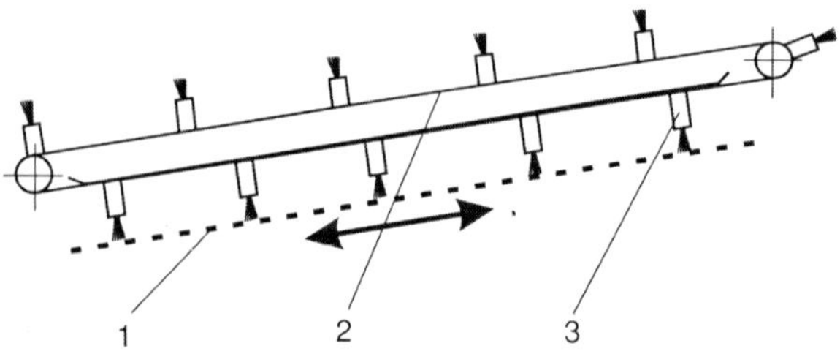

FIGURE 4.63. Sieve cleaning device with brushes. 1–sieve, 2–side belts, 3–laths with brushes.

Where:

G_O: weight of an object; F_R: flow resistance in N; C: overall drag coefficient; A_p: perpendicular cross section of the body-object in m^2; ρ_f: fluid density in kgm^{-3}; and v: body-object critical or terminal velocity in ms^{-1}.

The term used is *overall drag coefficient* that contains also *frictional drag coefficient,* which can be neglected for laminar flow (Mohsenin, 1980). There are some diverse apparatuses available for the measurement of critical velocity, of which the vertical glass duct is the one used mostly (Grochowicz, 1980). Many agricultural materials contain an admixture after harvesting with significantly different airflow characteristics. The typical ones are chaff in cereals, and sand or stalks in herbaceous MAP. However, it is not simple to perform successful separation and classification based on the difference in airflow characteristics—critical velocity. It is important to have a proper dosing of material. If the dosing of material is improper, some particles could be in the flow shadow and be incorrectly classified. Airflow has a mostly inhomogeneous field that can also cause improper classification. An example of airflow separation is shown in Figure 4.7c, for fresh cut plant material. The material is thrown into the airflow and carried away by the flow, further or closer, depending on the airflow characteristic. This type of separator is shown in Figure 4.64a. The device with airflow lifting of lighter particles, with lower critical velocity, that is, particles with higher airflow resistance is shown in Figure 4.64b. Both devices are simple and inexpensive and many diverse models are used. The type shown in Figure 4.64a is used more often, for example, combine harvester cleaning fan. The demerit of both types is the low accuracy of separation and classification due to the inhomogeneous airflow velocity field. The second problem is the realization of uniform dozing. Depending on the process demands, these devices could be used successfully. Sometimes multipassing of material is practiced to provide successful separation and classification of the desirable component. An example of a device where airflow is lifting "easier" particles, Figure 4.64b, is schematically shown in Figure 4.65.

By the opening and closing of the sliding gate (7), airflow intensity is regulated. The Plexiglas window (12), enables visual control of the separation and classification effects, for example, if all stalks drop down and all leaves fly up, then the separation is proper. If the separation is not proper, the airflow should be reduced or intensified using the sliding gate (7). Vibro-dozer (2) should bring material to the vertical duct (3) in as uniform and as loose a layer as possible. The "easiest" material will be caught in the cyclone (4). Small and medium producers can effectively use this facility; it showed especially good results in the separation of stones and sand. Such an airflow separator is shown in Figure 4.66. A simple and innovative airflow separator is schematically shown in Figure 4.67.

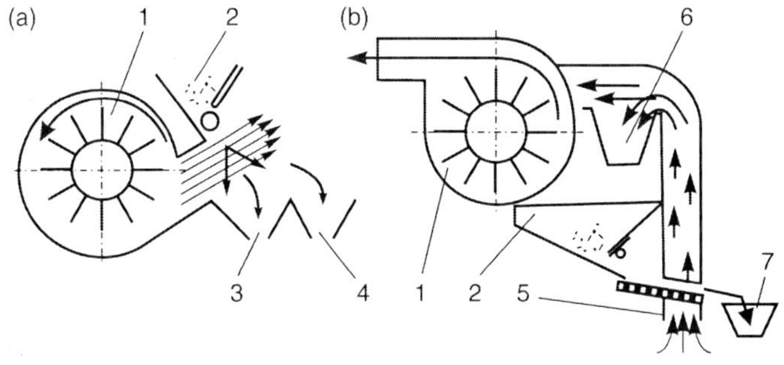

FIGURE 4.64. Simple devices for airflow separation and classification. (a) with throwing out of particles, (b) with particles taking off; 1–ventilator, 2–hopper with dozer, 3–"heavier" material output, 4–"easier" material output, 5–vertical duct with sieve, 6–"lighter" material hopper, 7–"heavier" material hopper.

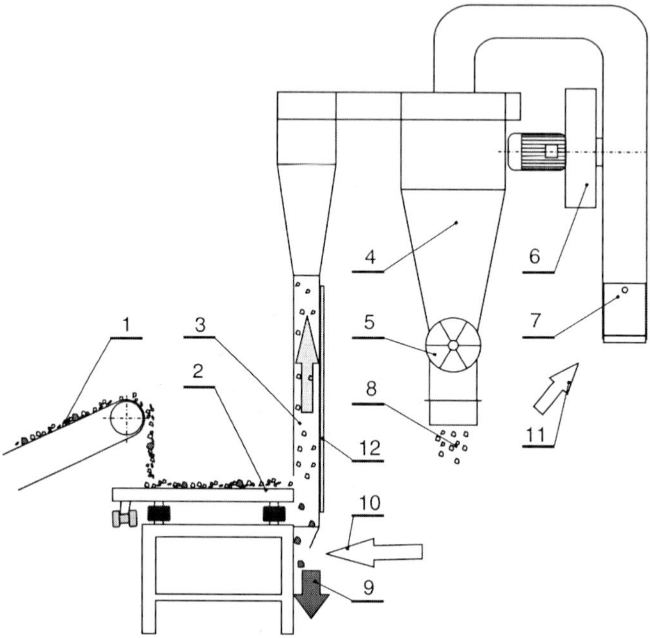

FIGURE 4.65. Vertical duct airflow separator. 1–feeding band, 2–vibro-dozer, 3–vertical duct, 4–cyclone, 5–cell dozer, 6–ventilator, 7–air flow regulator, slider, 8–"light" material, 9–"heavy" material, 10–main air inlet, 11–secondary air inlet, 12–Plexiglas window.

The difference in airflow resistance, that is, the ratio between surface and weight, is combined here with the centrifugal force generated by the rotation of the grate drum. Leaves with lower critical velocity and higher airflow resistance, i.e., the leaves that are longer, connected to the drum surface are disposed to the right positioned hopper (4). At the same time the airflow removes the "light" dust. Repeating of the processing steps can improve airflow separation and classification. It means that the material will pass more than once the zone where particles with different critical velocity are separated, for example, the cascade separator shown in Figure 4.68.

This separator is equipped with four cascades (3), that is, four separation zones. Part of the "lighter" particles will be carried out with the airflow to the top cascade. Some "light" particles still fall and

FIGURE 4.66. Vertical duct airflow separator.

slide down mixed with "heavier" material. The sliding and falling of material causes its loosening. At every further cascade the result of the separation is improved. The so-called zigzag separator is used more frequently. The vertical duct is designed to enable multiple cascades and separation zones, Figure 4.69. That enables multicorrection of incorrect separation as a result of lower velocity close to the duct walls. "Lighter" particles fall down in the middle of the "full" air velocity zone.

The behavior of grouped and disintegrated–loosened material—in the zigzag section is presented in Figure 4.70. There is a positive effect of loosening, the falling down of grouped material, which enables better separation in the lower sections.

Zigzag separators are very tall, approximately 5 m, because a certain number of sections are needed. The demerit of this facility is the particle size reduction, that is, the crushing, caused in cell dozers. The second demerit is its high price, which is an obstacle for purchase by small and medium-sized companies. Sometimes airflow

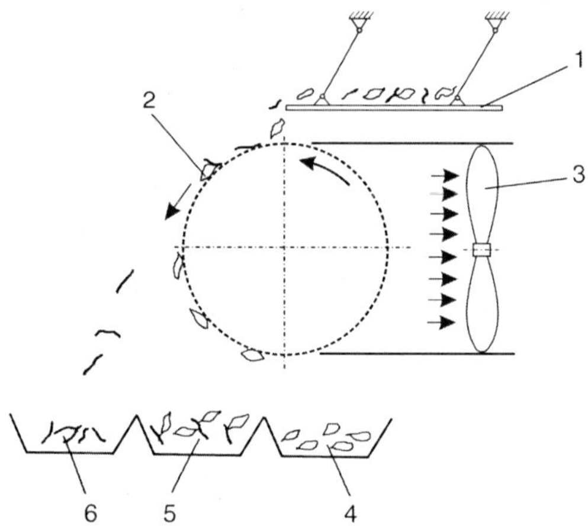

FIGURE 4.67. Device for airflow separation with perforated drum. 1–vibrating feeder-dozer, 2–perforated drum, 3–ventilator, 4–hopper for leaves, 5–hopper for leaves with stems, 6–hopper for stems. *Source:* Oebel, 1989.

separation can be combined with sieving. In that case, a typical selector, as shown in Figure 4.58 cannot be used because the desirable parts of herbaceous MAP are "lighter" than the residues, that is, stalks. They are usable only for the processing of grains and some fruits.

Separation and Classification According to the Difference of Coefficient of Friction and Rolling Resistance

Differences in the coefficients of friction of plant material can be used to separate undesirable/desirable parts. For some seeds, the difference in the friction coefficient is used to separate parts of leaves, chaffs, split stalks, stems, and the like. Some very simple devices could be used, for example, the spiral gravity system as shown in Figure 4.71. The parts with lower coefficient of friction gain higher velocities while sliding over the spiral, and are thrown outside, toward the box (1). The particles with lower velocity, that is, the particles with higher coefficient of friction, fall into the box (2). The only demerit of this simple device is its low output.

There are diverse possibilities of utilizing the difference of rolling resistance for material separation and classification. Grains typically have lower rolling resistance than straw or parts of straw. This prop-

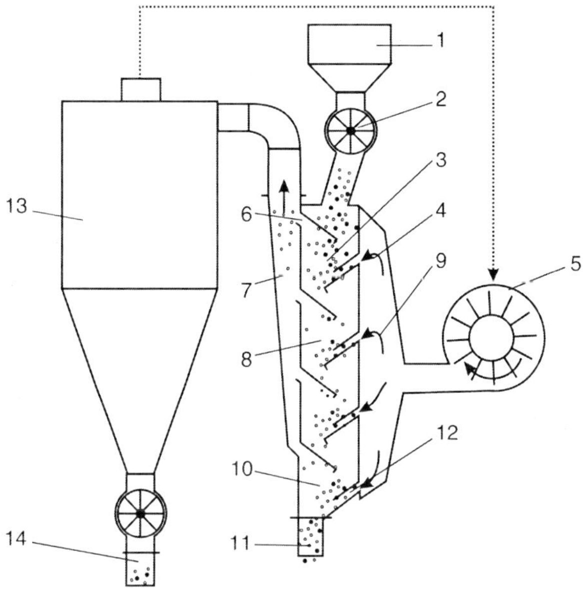

FIGURE 4.68. Cascade airflow separator. 1–hopper, 2–cell dozer, 3–cascade, 4–upper air intake, 5–ventilator, 6–upper air outtake, 7–vertical duct, 8–central air outtake, 9–central air intake, 10–lower air outtake, 11–"heavy" material outtake, 12–lower air intake, 13–cyclone, 14–"light" material outtake.

erty is widely used for cleaning of seed material. The scheme of two devices that use this principle is shown in Figure 4.72.

Here, the selection of band material is also important, but less than in the case of separation and classification according to the difference of coefficient of friction. The device shown in Figure 4.72b offers the possibility of classification into more than two groups.

The devices for separation and classification according to the coefficient of friction and rolling resistance can be used as a part of the processing lines or even as a part of a single machine. Sometimes sliders or conveyors used for material transport are also designed for these purposes.

Other Separation and Classification Procedures

For the processing of some MAP, combined separators could be used. A typical example is the gravity separator, widely used for the

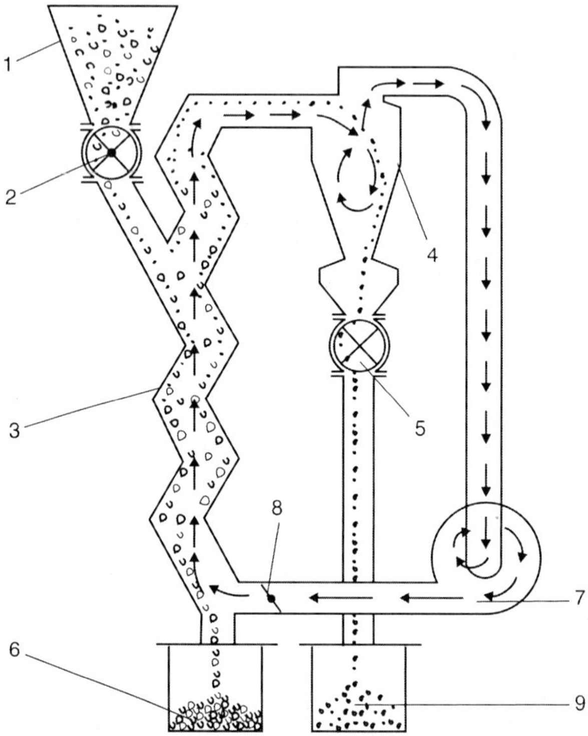

FIGURE 4.69. Zigzag airflow separator. 1– hopper, 2–dozer with cells, 3–zigzag vertical duct, 4–cyclone, 5–dozer with cells, 6–"heavy" material bin, 7–ventilator, 8–air flow regulator, 9–"light" material bin.

processing of seeds. Separation is here based on the difference in airflow characteristics, rolling resistance, density, and even coefficient of friction. An example of this separator is shown in Figure 4.73.

The working plane-grate (1) is inclined longitudinally, at angle β, and perpendicularly, at angle α. These slopes can be adjusted. The intensity of the airflow and oscillations *(e)* are also adjustable. The combined effects of airflow, oscillations, and sliding generate separation of material of diverse weight. Lighter material forms an upper layer that moves down, and heavier material forms a lower layer that moves up. Many reputable companies, suppliers of cereals and seed producers, produce this type of separator. It is only due to the rela-

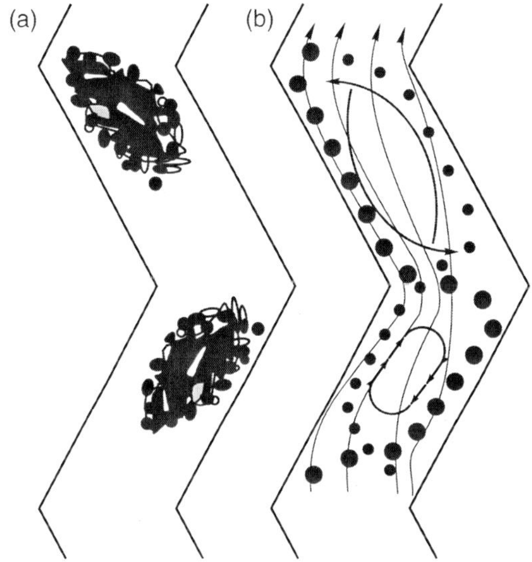

FIGURE 4.70. Material flow in the zigzag duct. (a) grouped material; (b) loose material. *Source:* Kabelitz, 1997. Reprinted with permission.

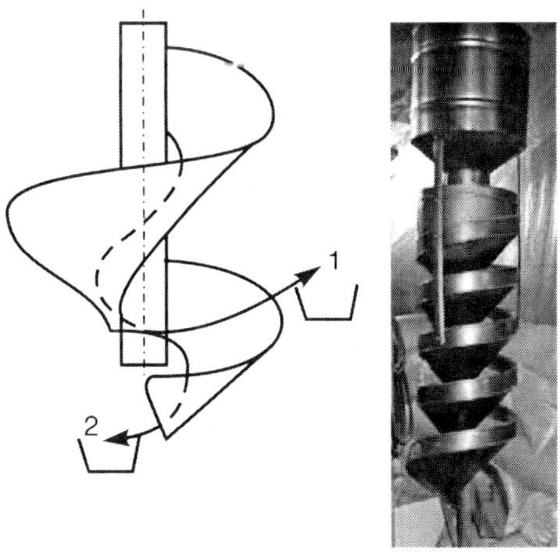

FIGURE 4.71. Spiral gravity system for separation according to difference in coefficient of friction. 1–outlet for lower coefficient of friction parts, 2–outlet for higher coefficient of friction parts.

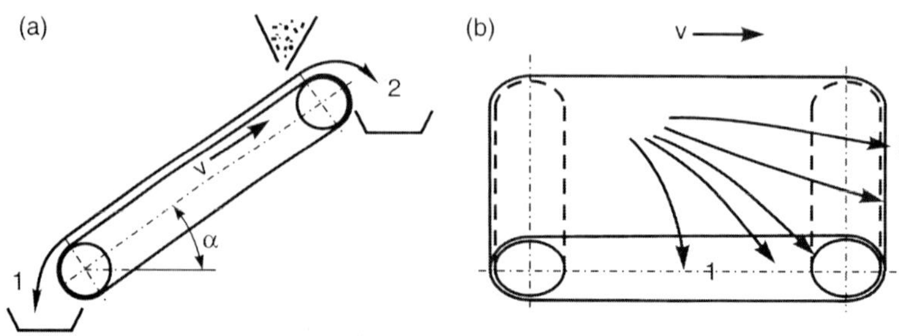

FIGURE 4.72. Devices for separation and classification based on rolling resistance difference. (a) longitudinally inclined band conveyor, (b) perpendicularly inclined band conveyor: 1–outlet for particles with lower rolling resistance, 2–outlet for particles with higher rolling resistance.

tively high price that this device is not widely used by small and medium-sized companies processing MAP.

For dust removal, electric separators are primarily used, for example, the device shown in Figure 4.74. The operating principle is based on the difference in the electric properties of the plant material and dust. These facilities are commonly used for the processing of cereals and flour, and rarely for the processing of MAP. Some larger processing lines are equipped with devices for the separation of stones and metal parts. That is needed if wild crafted material is processed and if some stone-metal sensitive parts are part of a processing line, for example, a cutter. Diverse airflow separators are widely used for these purposes, for example, the vertical duct facility using the difference in material density presented in Figure 4.65 and Figure 4.66. For metal parts removal, ducts with a permanent magnet or electromagnet could be used. The material flow is reduced to a thin layer to enable all material to cross the magnetic field.

Separation based on the difference in electrical properties has been commonly used for seed cleaning. This procedure is based on the difference in the electrical conductivity of seeds and admixtures, for example, dust. The bodies with lower electrical conductivity are able to hold electrostatic charge longer, after passing through an electrical field (Mohsenin, 1980). Schemes of two devices for separation based on difference in electrical properties are shown in Figure 4.74.

The left device has a rotating drum (4) with positive charge, which attracts material crossing the negatively charged field of electrode (3).

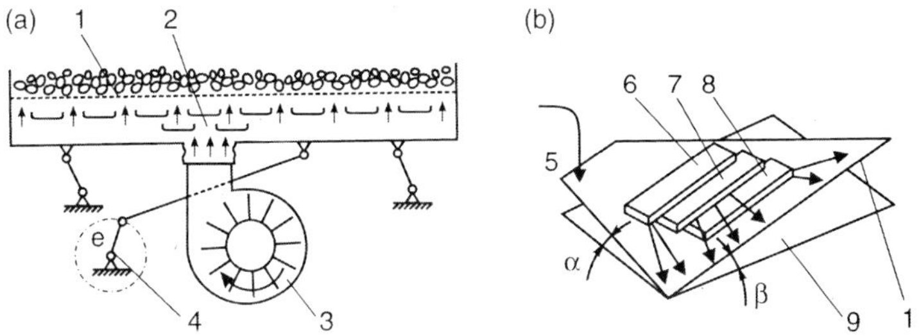

FIGURE 4.73. Gravity separator. 1–working plane-grate, 2–airflow distributors, 3–ventilator, 4–eccentric actuator for oscillation generation, 5–material intake, 6–upper layer of "lighter" material, 7–middle layer, 8–lower layer of heavier material, 9–horizontal plane.

The particles with higher electrical conductivity fall down to the left, due to the weight and centrifugal force. The others, with lower electrical conductivity, are in contact with the drum longer and fall down to the right. Adjustable dividers (6) enable precise separation of diverse materials. The device presented at the right has a similar working principle, but here the particles fall down, and the trajectory of the free fall depends on their electrical charge—negatively charged box wall attracts to the left only positively charged particles. The facilities for the separation of material based on the difference in electrical properties are mostly used for dust separation in cereals and flour processing plants but could be used for MAP as well. Other efficient procedures, for the separation and classification of MAP based on the difference of other physical and chemical properties of material, are still either in the phase of development or too expensive for wider use.

Other Types of Mechanical Processing

There are many additional diverse processing procedures practiced in accordance with the type of the final product. Manual inspection, as a final phase of process, is practiced in many small and medium-sized companies in the developing countries. The material is distributed in a thin layer on the table, or better, on an inspection band and observed. Experienced persons are making a final separation of

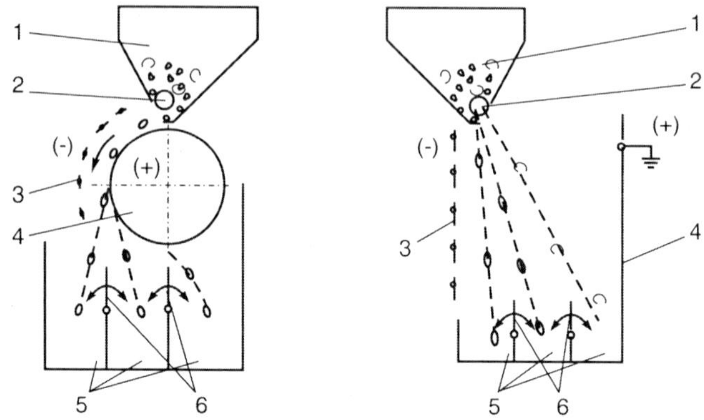

FIGURE 4.74. Facilities for separation based on difference in electric properties. 1–hopper, 2–dozer, 3–negative charged body–electrode, 4–positive charged drum i.e., box, 5–selected material departments, 6–adjustable dividers.

all undesirable admixtures. Packing of MAP for storage or transport is a common operation. Big sacks made of natural or synthetic fibers are used. For the filling of the sacks, diverse auxiliary tools are used, for example, a very simple self-made frame shown in Figure 4.75. A big diameter ring enables easy filling of plant material without pressing.

Medium and large-sized companies use industrial filling devices, with continuous or semicontinuous weighing. If the plants are not sensitive to mechanical damage, pressing is applied. Some presses aimed at other purposes are used, for example, for paper, but presses are also exclusively developed for MAP. Pressed material is tied using plastic bands commonly utilized for packaging and put into textile sacks. This enables safe storage and transport.

Many small and medium enterprises are interested in distributing part of their products directly to the market at the local level by making mixtures of tea or spices. For these purposes diverse mixing devices are used. One of the typical devices is a barrel made of stainless steel with a diagonal shaft, Figure 4.76. This barrel is rotated manually. Changing of the material path causes sufficient mixing. The problem is proper cleaning of the barrel, that is, preventing the residues of the previous mixture from being mixed with the next one. The residues accumulate mostly at the acute angle corners. This can be solved by designing a special vessel with side angles larger than 90°.

FIGURE 4.75. Simple self-made frame, auxiliary equipment for filling sacks.

Examples are shown in Figure 4.76 and Figure 4.77. An alternative for rotating vessels is a mixer with two horizontal counter-direction augers. This type can be used only for mixtures where the components are not sensitive to the intensive mechanical processing or rubbing. On the other hand, the particle size will be reduced and an increased content of undersized particles is possible. That can provoke low quality of mixture or even unusable mixture.

There are two types of tea packaging for market—bags and filter bags. The bags contain 50 to 100 g of herbs or mixture of herbs. In some cases, the bags are bigger, up to 1 kg, if they are aimed for large-scale consumers, for example, hospitals, schools, and so on. For the filling of the bags, volumetric dosing–metering devices with additional manual correction on balance are used. The bags are then closed and put in appropriate packages, containing all data needed to

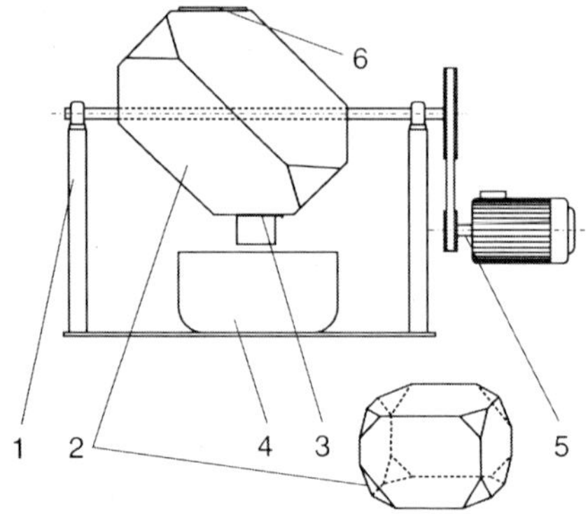

FIGURE 4.76. MAP mixing device–barrel. 1–frame, 2–mixing box, 3–outlet tube with slider, 4–hopper, 5–drive, 6–filling opening with cover.

FIGURE 4.77. MAP mixing device with special vessel.

(a)

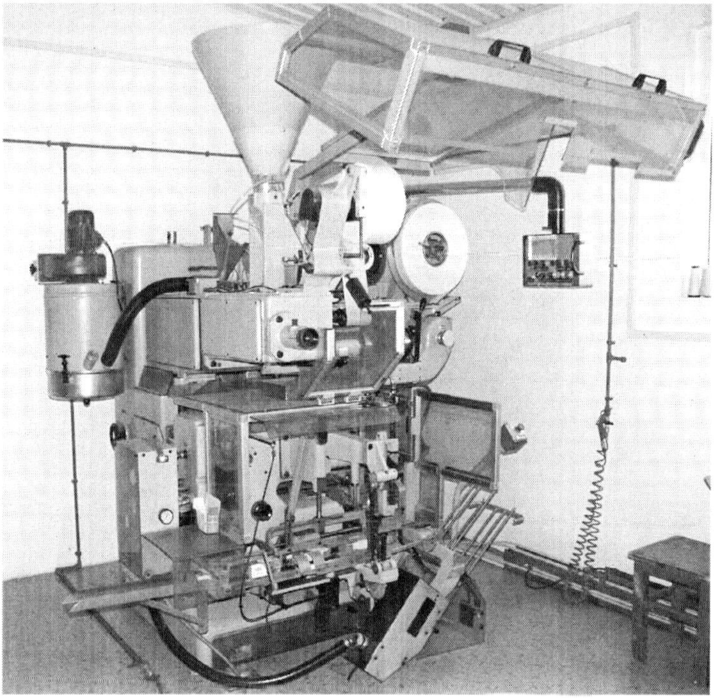

(b)

FIGURE 4.78. Filter tea bags making machines. (a) low capacity, (b) high capacity with four rows.

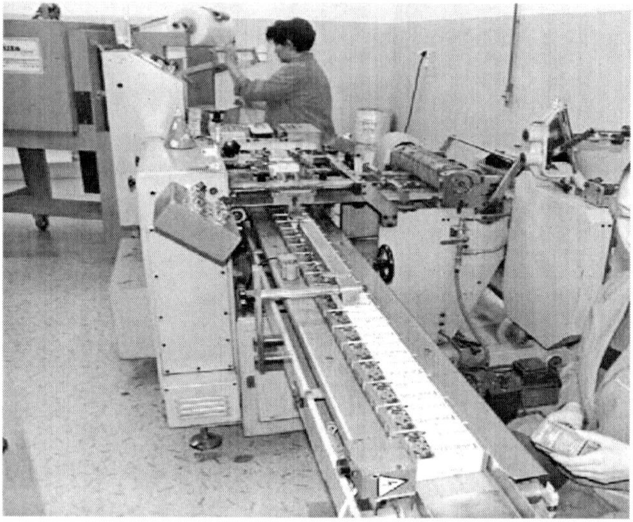

FIGURE 4.79. Closing and wrapping of boxes.

bring them to the market. This is the simplest procedure that provides market-ready product and it is practiced in many developing countries. The higher level of finalizing is the production of filter tea bags. There are diverse machines for this purpose. The simpler machines make tea bags without thread. An example of a filter bags packaging machine is given in Figure 4.78a. There are also high capacity machines that produce several rows of bags simultaneously, as shown in Figure 4.78b.

The closing of the tea boxes and the wrapping of the boxes with cellophane can be manual or mechanized. An example that illustrates these operations is presented in Figure 4.79.

REFERENCES

[Anonymous] 1993. ASAE S424.1 Method of determining and expressing particle size of chopped forage materials by screening. In ASAE Standards 1993. American Society of Agricultural Engineers, St. Joseph, MI: 459-461.

[Anonymous] 2003a. Herba-TEC products presentation. Available From: www.herba-tec.de.

[Anonymous] 2003. Prospects of company Europrima. Available from: www.europrima.co.yu.

Geyer M. 1999. Gemüsereinigung, KTBL–Kuratorium für Technik und Bauwesen in der Landwirtschaft, Darmstadt; 92 pp.

Grochowicz J. 1980. Machines for Cleaning and Sorting of Seeds. US Department of Agriculture, National Sciences Foundation, Washington DC, Foreign Scientific Publications Department, National Center for Sciences, Warsaw; 427 pp.

Heindl A. 2003. Probleme und Lösungsmöglichkeiten bei der Aufbereitung und Trocknung von Arznei- und Gewürzpflanzen. *Zeitschrift für Arznei und Gewürz Pflanzen;* (8)1: 33-36.

Kabelitz L. 1997. Korrekturmaßnahmen bei Qualitätsmängeln von Drogen. *Arznei-& Gewürzpflanzen;* (2)3: 120-125.

Kaltschmitt M, Hartmann H. 2001. *Energie aus Biomasse,* Springer-Verlag, Berlin Heidelberg, New York: 770 pp.

Konstantinovic M. 2000. Design, Construction and Testing of Dog Roses Crusher, MS Thesis, Faculty of Engineering, Novi Sad: 81 pp.

Lacak R. 1998. Design, Construction and Testing of Chamomile Separator, MS Thesis, Faculty of Engineering, Novi Sad: 53 pp.

Martinov M, Pejak M, Veselinov B. 1991. Development of machine for leaves removal of dry peppermint. XV Symposium of Yugoslav Society of Agricultural Engineering, Opatija, Book of Proceedings: 178-184.

Martinov M, Adamovic D, Veselinov B. 1992. Development of machines for leaves removal of fresh peppermint. XVIII scientific meeting of Vojvodina Society of Agricultural Engineering, Donji Milanovac, Book of Proceedings: 151-156.

Mohsenin NN. 1980. *Physical Properties of Plant and Animal Materials.* Gordon and Breach Science Publishers, New York: 742 pp.

Müller J. 2000. Nacherntetechnologie–Arzneipflanzen. Teaching material, University of Hohenheim, Institute for Agricultural Engineering in Tropics and Subtropics, Stuttgart.

Mulugeta E, Geyer M, Oberbarnscheidt B. 2003. Düsen für die Gemüsenwäsche. *Landtechnik;* (58)4: 256-257.

Oebel H. 1989. Verfahrenstechnik von Heil- und Gewürzpflanzen, MS Thesis. Rheinischen Friedrich–Wilchelm Universität Bonn, Institut für Landtechnik, Bonn: 125 pp.

Sijacic S. 1999. Processing and cutting of root MAP–example marshmallow, MS Thesis. Faculty of Engineering, Novi Sad: 69 pp.

Tesic M. 1984. Principi rada masina za zetvu travnatih materijala (*Engineering Principles of Hay and Forage Harvesting Materials*). Faculty of Engineering, Novi Sad: 240 pp.

Veselinov B. 2003. Impact of different procedures of peppermint size reduction to the quality of plant material. PhD Thesis, Faculty of Engineering, Novi Sad: 110 pp.

Chapter 5

Extraction

Yurtsever Soysal
Serdar Öztekin

INTRODUCTION

Most plants synthesize numerous active ingredients that accumulate within various organs during their development stages. It is thought that these natural compounds are nonessential for the basic photosynthetic or respiratory metabolism, but play a critical role in the fitness and survival of the species to defend against fungi, bacteria, insects, and viruses, and to attract insects for pollination. Such compounds are known as secondary plant metabolites, which can be extracted; some are termed essential oils. Essential oils are the aromatic portions of the plants located within distinctive oil cells, glands, or internal secretor canals. In some exceptional cases, essential oils are formed during the extraction of the source plant material. However, some essential oils are formed only as a result of enzymatic reaction. They originate from a single botanical source and can be described as highly concentrated, volatile, and aromatic essences of the plant. Each essential oil contains hundreds of organic constituents that are responsible for their therapeutic actions and the characteristic odor of the source plant material. These compounds are classified as monoterpenes, sesquiterpernes, aldehydes, esters, alcohols, phenols, ketones, oxides, and coumarins. Essential oils are extracted from various parts of the plant like leaves, roots, wood, bark, seeds/fruits, flowers, buds, branches, twigs, or whole plants. About 65 percent of the

Medicinal and Aromatic Crops

doi:10.1300/5895_05

essential oils produced in the world are obtained from the woody plants that are trees and bushes (Baser, 1999).

Essential oils and other flavoring products are isolated from the plant material by various extraction techniques such as cold pressing or expression, distillation, solvent extraction, vacuum microwave distillation, maceration, and enfleurage. The products of extraction are usually termed as concretes, absolutes, pomades, and resinoids, which are not regarded as essential oils. The term *essential oil* is only used for distilled or extracted oils. Although the extraction and distillation refer to different isolating techniques, in this text, we use the term *extraction* in a broader sense to define the process of obtaining extractable compounds like essential oils and other flavoring products from the plant materials (medicinal and aromatic plants), mixtures and compounds by chemical, physical, or mechanical ways.

Essential oils are generally obtained by various distillation techniques such as water distillation, water and steam distillation, steam distillation, and expression or cold pressing. Among them steam distillation or hydrodistillation is the most widely accepted method for the production of essential oils on a commercial scale. Approximately 90 percent of the essential oils are produced in this way (Runham, 1996). Distillation is a heat-dependent separation and purification process of a liquid mixture based on the vapor pressure difference. It can be defined as a method of extracting an essential oil from the plant material by evaporation and subsequent condensation of a liquid. On the other hand, flavoring products such as concretes, absolutes, resinoids, pomades, tinctures, extracts, and oleoresins are produced by extraction with organic solvents or liquefied gasses, such as solvent extraction and supercritical fluid extraction.

It is well known that the composition of the finished products is highly dependent on the method used to extract them. Naturally, the most valuable products require the most sophisticated processing techniques (Chung, 2000). Further processing causes both the commercial value and the therapeutic potent of the finished products to exhibit an increasing tendency because of the increasing concentration of the end product and sophistication in processing, Figure 5.1. Similarly, the cost of production depends highly on the scale of the production small, medium or industrial scale production, raw materials, processing equipment, energy, and labor requirement.

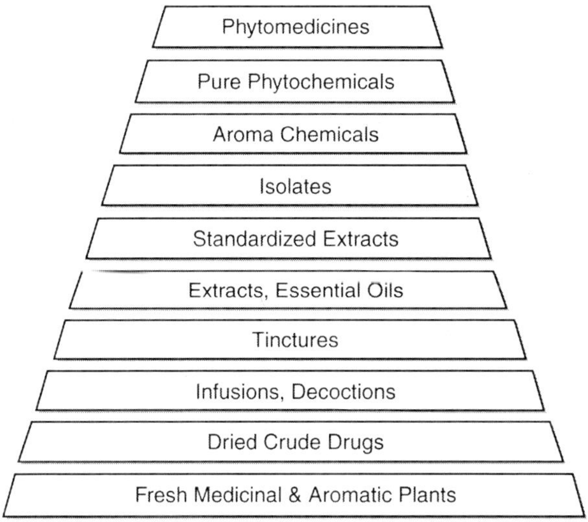

FIGURE 5.1. Progressive stages of MAP processing and the value of the end products. *Source:* Adapted from Chung, 2000.

CONVENTIONAL EXTRACTION TECHNIQUES

Distillation

The process of distillation is an age old essential oil extraction method whereby the liquid is separated into components that differ in their boiling points. The distillation technology is relatively simple and adoptable in rural areas. Basically, a simple distillation unit consists of four parts: furnace (heat source), distillation still, condenser, and oil separator, Figure 5.2.

Although the principle of distillation is the same, practical application of this technique differs depending on the type and physical form of plant material. Three types of distillation systems are used in practice. These are water distillation (distillation with water), water and steam distillation, and steam distillation (distillation with steam). The yield, composition, quality, and commercial value of essential oils are greatly affected by the type and efficiency of the distillation unit, as well as the age of the harvested plant material and ecological conditions where the plant material is cultivated or wild harvested.

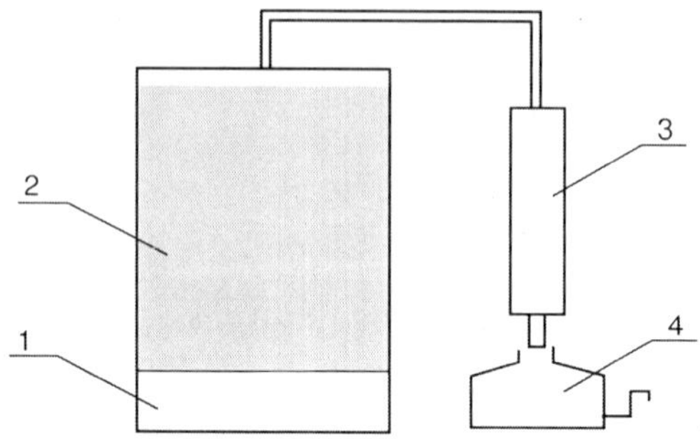

FIGURE 5.2. Simplified plan of the water distillation plant. 1–furnace, 2–distillation still, 3–condenser, 4–oil separator.

Water Distillation

Water distillation is the oldest and the cheapest distillation method which is simple in design and easy to construct. It is generally used to extract the essential oils of the dried or powdered plant material—that is, spice powders, ground woody plants like cinnamon bark—as well as some flowers such as rose and orange and very tough materials such as roots, woods, or nuts.

The simplified flow diagram of the process of water distillation is given in Figure 5.3. In this method, the plant material is filled into the still containing a certain amount of water to ensure they are completely immersed and soaked. The water is then boiled by direct heating of the distillation vessel containing the plant material and water suspension. Depending on the product characteristics improved distillation control can be obtained at reduced or excess pressure. The distillation still can be heated by a steam jacket. During this boiling process the volatile component, essential oils from special structures inside or at the surface of the plant material, is mostly extracted at temperatures just below 100°C by diffusion mechanism. The extracted volatile constituent is transported along with the steam to the condenser where the mixture of vapor is cooled and transferred to the liquid phase. This mixture is then taken to the oil separator called the Florentine vessel. Due to the density difference, the essential oil is decanted from the water. The remaining distillate, which is a byproduct of distillation called floral water or hydrosol, can be added to another distillate and redistilled together with it.

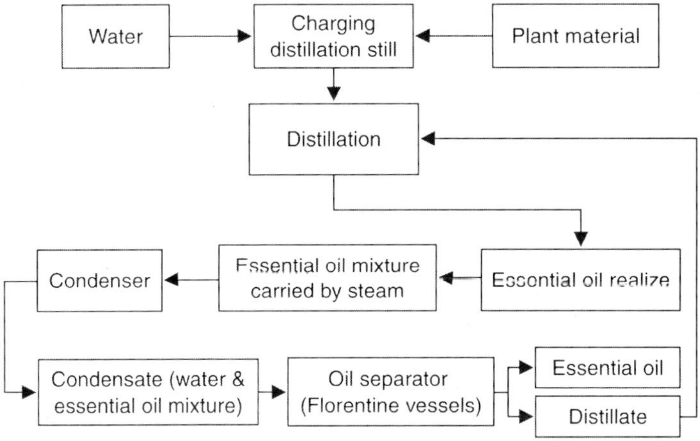

FIGURE 5.3. The flow diagram of the water distillation process. *Source:* Adapted from Lawrence, 1995.

The main disadvantage of this method is that the plant material to be distilled is always in contact with the surrounding boiling water which will be used to create steam for distillation. To prevent the agglomeration of dense plant material and to ensure that the plant material is always in contact with boiling water, the plant material should be stirred throughout the distillation process. Otherwise, as the agglomerated dense plant material will settle down at the bottom of the still, these material will be overheated and thermally degraded, which leads to the formation of off-notes. On the other hand, as the distillation still is directly heated by a furnace, water in this still must always be more than enough to last throughout the distillation process to prevent the overheating and charring of the plant material which leads also to the formation of off-notes (Lawrence, 1995). Further disadvantages of water distillation are as follows (Lawrence, 1995; Vukic et al., 1995):

- energy becomes expensive due to the slower process
- produces lower quality oil
- incomplete release of the essential oil
- requires skilled labor and large number of stills as charge has to be less dense

However, commercial applications of this method can be observed in some developing countries like India, Turkey, and Sri Lanka for the distillation of cinnamon bark, sage, rose, and other flowers. A certain amount of rose oil in rural Turkey is produced by water distillation of fresh rose petals. As an example for water distillation process, the production steps of traditional and industrial rose oil are described in the following paragraphs.

Traditional Rose Oil Production

The rose harvesting season usually begins in May and continues until the end of June. To obtain high quantity and quality rose oil, the blossomed mature roses should be picked early in the morning. Hand harvesting is generally done by a member of the producer family, mostly by women, Figure 5.4. Harvested rose flowers are filled into gunny sacks and transported to the small distillery barn, Figure 5.5. Traditional rose oil distillation is carried out in tinned copper or galvanized steel open fired stills of 120 to 300 liters capacity. These stills have a spherical removable head that is connected to the pipe of the cooling unit-condenser. Lukewarm pool water is generally used as a cooling fluid. The spherical head works as a primary cooling unit by returning the phlegm back into the still and liberates rose oil more effectively. For each run approximately 10 to 12 kg of freshly harvested rose flowers and 50 to 60 liters of water are filled into the distillation still, Figure 5.6. After carefully stirring the mixture to ensure the rose flowers are completely immersed and soaked, the removable head of the still is tightly sealed with wet clay and the pipe of the cooling unit is connected to the head of the distillation still, Figure 5.7. When the water in the still boils, the steam with rose essential oil is carried through the top of the distiller into the condenser where the steam (water vapor and the essential oil mixture) is then condensed. After the condensation, water and essential oil are collected in a ca. 9 liter glass collection flask for passive separation. This flask is replaced by another flask after collecting approximately 5 to 7 liters of the first distillate. In general, for each run, 14 to 16 liters distillate is obtained within 1.5 to 2 h. The rose oil from this distillate is then filtered and stored until all the roses for the season have been distilled, while the separated waters are collected for redistillation, Figure 5.8.

FIGURE 5.4. Hand harvesting of rose flowers.

FIGURE 5.5. Charging distillation still with freshly harvested rose flowers. Outside view of barn *(top left)*.

At the end of the season, all the essential oils obtained from different charges are blended to obtain the final rose essential oil.

Industrial Rose Essential Oil Production

For large-scale rose oil production freshly harvested mature rose flowers are filled into the gunny sacks and transported by trucks to

FIGURE 5.6. Charging distillation still with water.

FIGURE 5.7. Sealing head with wet clay and connecting the cooling pipe.

the distillery where they are weighed and loaded into the still body, Figures 5.9 and Figure 5.10. The lid of the still is tightly sealed. Typically, each distillation still has a charge size of about 500 kg fresh rose flowers and about 1,500 liters of fresh water. The rose and water mixture is heated with the steam jacket or steam coils inside the bot-

FIGURE 5.8. Filtering the rose oil.

tom of the still powered by a separate steam generator, Figure 5.11. The water in the still boils and the steam takes up the essential oil. The steam with rose oil vapors moves through the condenser usually cooled with running lukewarm pool water, where the mixture is then condensed, Figure 5.12.

The resulting mixture of condensed water and essential oil is then collected in Florentine vessels with a capacity of 200 liters, Figure 5.13. In general, the first distillation takes 1 to 1.5 h and produces about 25 to 30 g green oil or first oil, Figure 5.13; *bottom left.* After decanting this highly concentrated and most valuable first rose oil floats onto the Florentine vessel, the remaining distillates are pumped into the large tanks until the second distillation—redistillation—is performed, Figure 5.14. This second distillation is performed in a separate still. It takes about 1 to 1.5 h and produces yellow or second oil. Around 80 percent of the oil from the distillate is extracted by this process. These first and second distilled oils are then decanted and stored for marketing.

FIGURE 5.9. Transporting rose flowers to the distillery.

FIGURE 5.10. Filling the distillation still with rose flowers.

At the end of the season, all the essential oils, first and second oils, obtained from different lots are blended to obtain the final rose oil. The remaining floral water called rose water is also a valuable product. Depending on the season and the quality of rose flowers, 3 to 5 tons of flowers produce 1 kg of rose essential oil and 500 kg of floral water.

FIGURE 5.11. Fuel oil fired separate steam generator.

FIGURE 5.12. A view from steam distillation unit.

FIGURE 5.13. First rose essential oil collected in the Florentine flask. First oil *(bottom left).*

FIGURE 5.14. General view of distillery. Two pieces of distillate collection tanks for second distillation *(top right).*

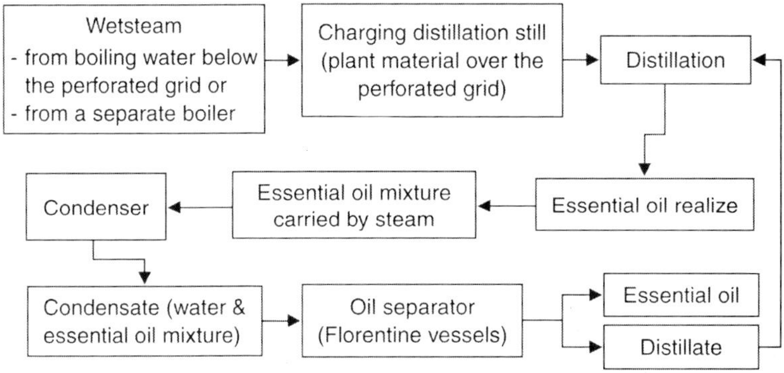

FIGURE 5.15. The flow diagram of the water and steam distillation process. *Source:* Adapted from Lawrence, 1995.

Water-Steam Distillation

Water and steam distillation is a combination of water distillation and steam distillation processes. It is mostly used for the distillation of fresh and dried plant materials such as citronella, peppermint, patchouli, lemongrass, eucalyptus, cinnamon leaf, clove, and thyme. The design of the equipment used is generally very similar to that used in water distillation (Lawrence, 1995).

A simplified flow diagram of the process of water and steam distillation is given in Figure 5.15. In this method, the plant material loaded into the distillation still is on a perforated grid to keep it above the water, which can be boiled by direct firing or by a steam jacket or a steam coil, and distilled with low pressure wet steam. The wet steam can be supplied from the boiling water below the perforated grid at the bottom of the distillation still or by a separate boiler. The plant material should be uniformly filled into the distillation still to provide a uniform steam passage (Vukic et al., 1995). Otherwise, the steam will pass through dominant channels and will be in contact with just some parts of the material. Further characteristics of this type of distillation can be summarized as follows (Lawrence, 1995; Vukic et al., 1995):

- Even if the plant material can not be in direct contact with the fire source beneath the still, as the walls of the still are good conductors of heat, unwanted still notes can also be obtained by overheating the plant material.
- The wet steam can make the lower plant material resting on the grid quite wet, and slow down the distillation process.
- Compared to the water distillation process,
  - it provides time and fuel economy and
  - higher yields of essential oil with minimal chemical changes.

Direct Steam Distillation

Steam distillation is the most common method of extracting essential oils on a commercial scale. Approximately 80 to 90 percent of the essential oils are produced in this way (Runham, 1996, Anonymous, 2003a). This method is used for fresh plant material that has high boiling point like seeds, roots, and wood. It is also used for fresh plant material such as peppermint and spearmint (Anonymous, 2003b), oil roses (Baser et al., 1990), and chamomile. The flow diagram of the steam distillation process and a simplified plan of the steam distillation plant are shown in Figures 5.16 and 5.17, respectively.

This type of distillation plant is similar to other types of distillation plants. The only difference is that there is no water inside the still body during distillation. After the plant material is loaded onto a perforated grid or in a basket or cartridge inside the still body, the saturated or sometimes overheated, pressurized dry steam that is generated in a separate boiler or a steam generator is injected below the plant material. As the steam will flow in the direction of less resistance, the plant material should be uniformly distributed in the still body. The essential oil of the plant material is extracted by the process of diffusion, while the steam passes through the plant material. The steam with a certain content of essential oil vapors moves through the condenser (Figure 5.18), usually cooled by running fresh water, where the mixture is condensed. The mixture of condensed water and essential oil is then collected and separated by decantation in Florentine vessels (Figure 5.19).

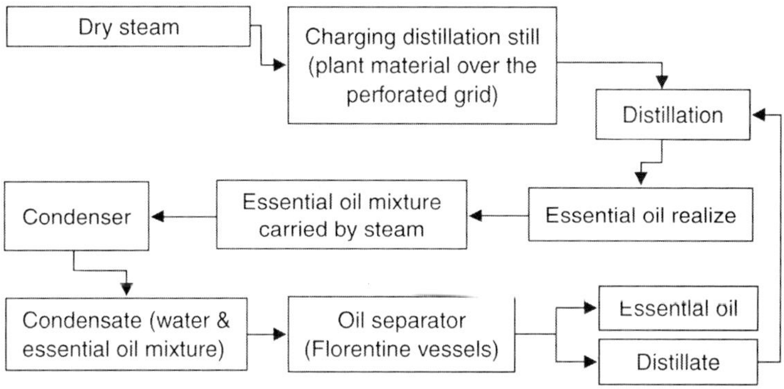

FIGURE 5.16. The flow diagram of the direct steam distillation process. *Source:* Adapted from Lawrence, 1995.

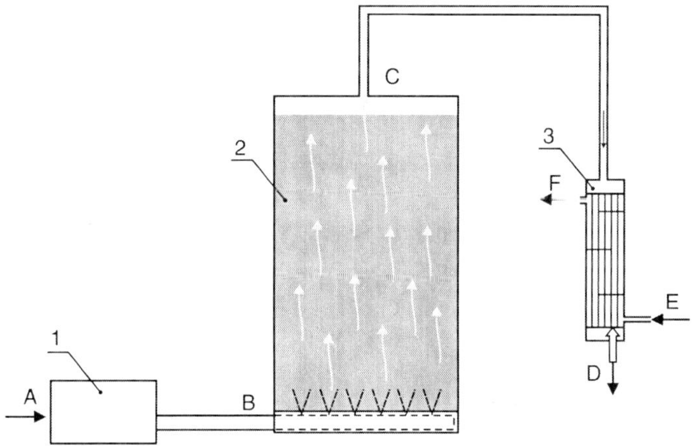

FIGURE 5.17. Simplified plan of the steam distillation plant. 1–steam generator, 2–still vessel, 3–condenser, A–fresh water point, B–dry steam point, C–mixture of water and essential oil vapors, D–condensed mixture of water and essential oil, E–fresh cooling water, F–used cooling water.

The process of steam distillation is more suitable for commercial scale operations. However, steam distillation has several advantages over the previously described variants (Lawrence, 1995; Vukic et al., 1995; FAO, 1992; Fraser and Whish, 1997):

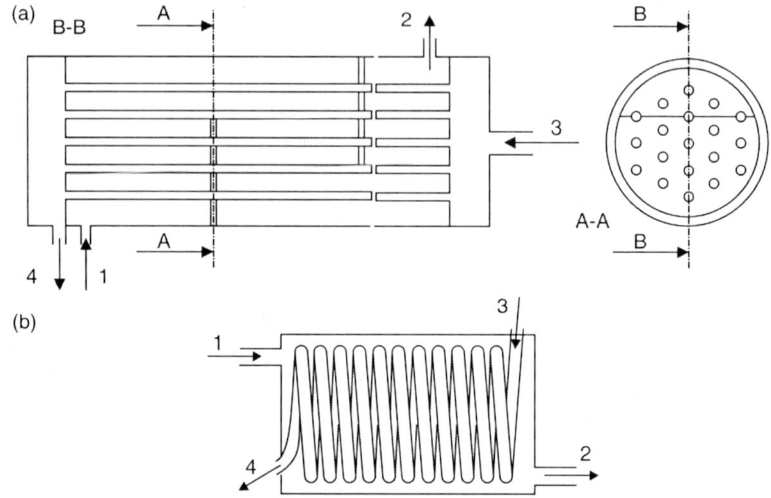

FIGURE 5.18. Main types of condensers. 1–cooling water inlet, 2–cooling water outlet, 3–distillate inlet, 4–condensate outlet (a) tube type, (b) spiral type. *Source:* Vukic et al., 1995.

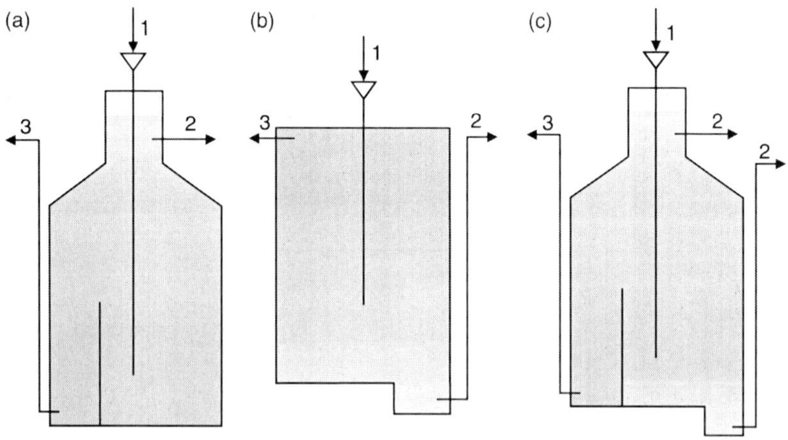

FIGURE 5.19. Florentine vessels. 1–condensate inlet, 2–essential oil outlet, 3–water outlet, (a) Florentine vessel for decanting of essential oils with lower density than water, (b) Florentine vessel for decanting of essential oils denser than water, and (c) Florentine vessel for decanting of both kinds of essential oils. *Source:* Vukic et al., 1995.

- It is more energy efficient.
- It is cost effective and is the cheapest method of essential oil extraction.
- It provides better control of the distillation rate.
- Quality of essential oils produced by this type of distillation units is more consistent and repeatable.
- There is the possibility of changing working pressure. That enables the use of low pressure for high volatile oils and higher pressures for low volatile ones.
- The rapid distillation is less likely to damage those oils containing reactive compounds like esters.

The obstacle for wider use of this type of distillation plant in developing countries and its biggest disadvantage is the high capital expenditure for the equipment (Vukic et al., 1995).

Many of the demands for the process of steam distillation are often contradictory. According to existing knowledge the most important demands are the following:

- Obtaining all, or most of the available essential oil from the raw material. To economically justify the process, approximately 85 percent of the essential oil in plant material must be obtained. The "dominant channels" can be formed in the still during water and steam and steam distillation processes. This happens because the plant material gets denser by absorbing water and fresh steam is now not able to move it, and pores in the randomly placed material at the beginning get larger. In this way only some part of the material is in touch with the steam and obtaining essential oil is far from complete. This can be prevented by using mixers such as an auger inside the still body or chopping of plant material before processing. The risk of losing considerable amount of essential oil exists while chopping the plant material.
- Realization of the process without any undesirable chemical transformation or degradation of the components of the essential oil.
- Reducing energy consumption and further requirements of the process. As the most efficient of all distillation processes, steam distillation is a very wasteful process. It is always necessary to

evaporate relatively high amounts of water to obtain a small quantity of essential oil. Water is never saturated with essential oil vapors. On the other hand, cooling water consumption is high. Cooling water temperature is usually 20°C at the input and is about 70°C at the exit from the condenser. The flow rate of this water is, for example, for processing of 300 kg of fresh plant material, up to 2,000 kg per hour. The problem is not just finding enough rich sources, but affording the price, and also the possible micro pollution with water at that temperature. Furthermore, water from the communal pipeline, for instance, has to be demineralized, prepared for use in steam generators and for some spiral type condensers (Vukic et al., 1995; Dachler and Pelzman, 1989).

- Increasing output of the processed plant material by achieving shorter processing time.
- The obtained components of the plant material should be, if possible, only the desirable essential oils.

Some improvements in steam distillation systems have been made for easier operation. In such a steam distillation system, mobile distillation unit, the harvested plant material is loaded into the mobile still body with a steam inlet system after wilting it and a subsequent chopping process (Lawrence, 1995). This mobile still body is then taken to the distillery where it is connected to the steam generator and condenser, Figure 5.20. Mobilizing the still body highly reduces the time for the whole distillation process and labor and consequently, increases the capacity. Some other improvements in the steam distillation process that will be discussed here are turbodistillation, hydrodiffusion, and continuous distillation.

Turbodistillation is becoming a popular alternative to steam distillation. Essentially, it is an improved version of the water and steam distillation process allowing faster extraction of essential oils from hard-to-extract plant materials such as bark, roots, and seeds. In this process, the plants soak in water and steam is circulated through this plant and water mixture. Throughout the entire process, the same water is continuously recycled through the plant material (Anonymous, 2003c).

Hydrodiffusion is another application of steam distillation whereby the steam at atmospheric pressure is passed into the plant material

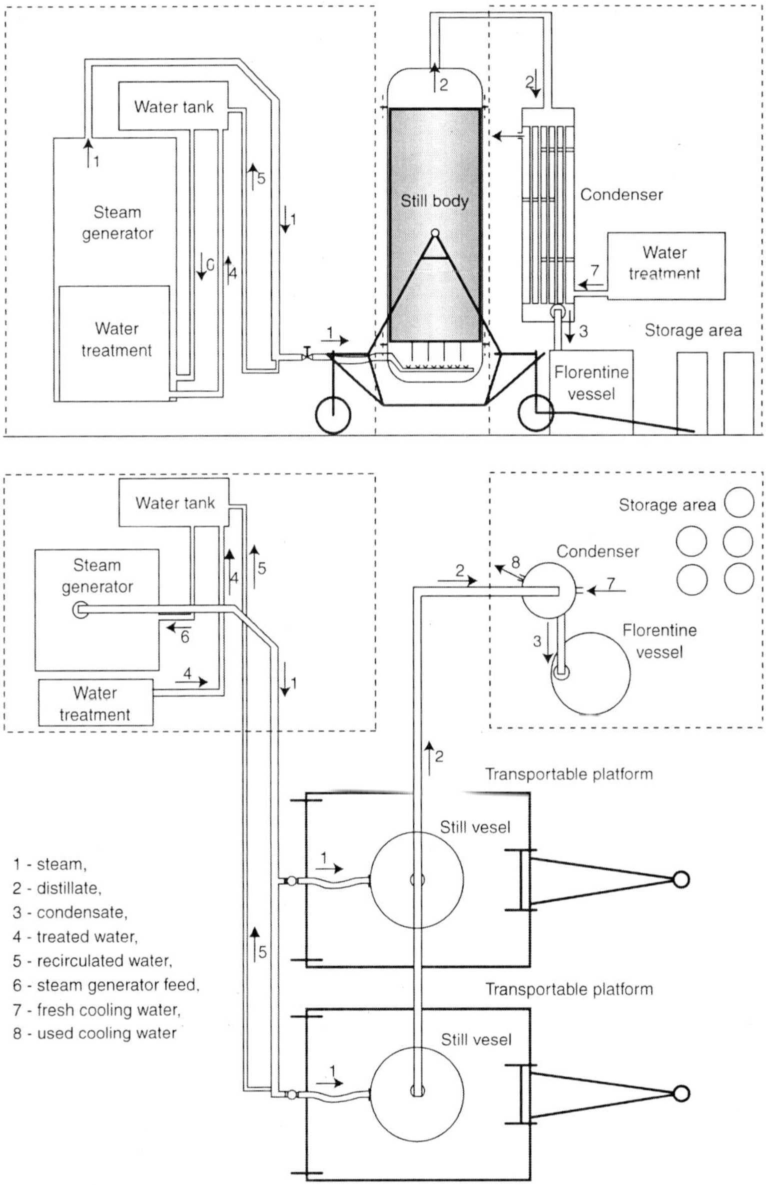

FIGURE 5.20. Mobile steam distillation plant. *Source:* Reprinted with permission from Rinder and Bomme, 1998.

from the top of the distillation still. In this process, for easier operation, the plant material is charged into the basket within the still and the tube type condenser is placed under the basket (Lawrence, 1995). In this way the steam can saturate the plants more evenly and in less time than with steam distillation. This method is also less harsh than steam distillation, and the resulting essential oils smell much more like the original plant (Anonymous, 2001a).

Several types of continuous steam distillation systems have been developed. Detailed descriptions of these systems are reported by Lawrence (1995). In these steam distillation units, the powdered plant material is generally conveyed by a pneumatic system or an endless screw into the distillation still body; within which it is brought into contact with superheated steam, and essential oil of the plant material is extracted.

Cold Pressing/Expression

Cold pressing or expression is a widely used method to extract the essential oils of citrus species such as bergamot, grapefruit, lemon, lime, mandarin, orange, and tangerine oils. Actually, a large amount of citrus oils are by-products of industrial processes (Simon, 1997). In this process, essential oil is obtained by scarification while the fruits roll over the sharp projections that penetrate the peel. This allows the piercing of essential oil containing sacs or glands located in the peel. Fruits are then cold pressed both for squeezing the juice from the pulp and for extracting the essential oils from the peel (Anonymous, 2003e). The realized essential oils during mechanical pressing under a spray of water are then trapped in an emulsion (Kunkar et al., 1993). This mixture is collected in a tank, filtered to remove the solid particles, and centrifuged to remove the essential oils from the water, cellulose, and pectin. The citrus essential oil extraction by scarification followed by cold pressing is now a mature extraction technology and is effectively used all over the world. The FMC Citrus Juice Extractor (FMC Corp.) and the Brown Peel Shaver (Automatic Machinery Corp.) are the most common types of commercially used extractors (Kunkar et al., 1993; Lawrence, 1995; Simon, 1997).

However, citrus oils produced by cold pressing or expression are not stable products because of the unsaturated terpene fraction ranging from 60 to 90 percent, depending on the species (Goto et al., 1997; Luque de Castro et al., 1999). They can be decomposed by heat, light, and oxygen resulting in low quality essential oils (Luque de Castro et al., 1999). Oxygenated aroma compounds in citrus oils (less than 5 percent) are mainly responsible for the flavor characteristics like grapefruit, lemon, orange, and lime (Sato et al., 1998; Luque de Castro et al., 1999). Therefore, to stabilize the product and to improve essential oil solubility, the terpenes that contribute only slightly to the main flavor characteristics should be removed by further processing called deterpenation or fractionation by using proper techniques like distillation, solvent extraction, or supercritical CO_2 extraction.

Solvent Extraction

Solvent extraction is the most widely used modern extraction method to obtain extracts from plant materials. In this process, various solvents are used to saturate the plant material and extract the aromatic compounds. Generally used solvents are the following:

- low boiling point organic solvents; propane, butane, hexane, methanol, ethanol, 2-propanol, acetone, methyl acetate, dichloromethane, dichloro-ethane, petroleum ether, toluene, and benzene
- water; boiled (decoctions), hot or superheated water (infusions)
- fats or waxes
- liquefied gaseous solvents; carbon dioxide, freons

In this process, nonvolatile components of the plant material like waxes, fatty oils, resins, and color pigments are also extracted due to the low selectivity of the extraction solvents. Therefore, solvent selection plays an important role in the extraction process to obtain the desired high quality products. There is no perfect solvent; each one has its drawbacks. The choice of solvent depends on several factors. For commercial use, the ideal solvents must have the following properties:

- high selectivity (does not dissolve the water content)
- low viscosity
- high solution capacity
- low latent heat of vaporization
- low boiling temperature
- stable (to heat, light, oxygen) and inert
- nontoxic and acceptable as food solvent
- readily available in large quantities
- recoverable and easily removable leaving no solvent residue
- nonflammable, inexpensive, and environmentally friendly

It is clear that all the demands for an ideal solvent are very difficult to achieve. In practice, only some of the demands could be realized.

Organic Solvent Extraction

The organic solvent extraction method uses several organic solvents to extract both essential oil and oleoresins that are then separated. The organic solvents used in this process are propane, butane, hexane, methanol, ethanol, 2-propanol, acetone, ethyl/methyl acetate, dichloromethane, dichloro-ethane, freons, petroleum ether, toluene, and benzene (Lawrence, 1995; Anonymous, 2003d). In this method, both dried and fresh plant materials are used for extraction. The fresh or dried plant material is filled into the extraction tank and then thoroughly mixed with a nonpolar solvent, such as hexane, until all of the aromatic compounds are completely dissolved in the solvents. The solvents dissolve the plant materials including essential oils, fats, waxes, and coloring pigments. After the solvent is saturated, the solvent is evaporated at reduced pressure. The product of the solvent extraction is generally a waxy product called a concrete (in the case of fresh plant materials) or resinoids (in the case of dried or dead plant materials). The dead organic plant material are balsams (e.g., benzoin), resins (e.g., amber), oleoresins (e.g., turpentine), and oleo gum resins (e.g., frankincense and myrrh) (Anonymous, 2004a). Resinoids range from liquids to solids. When the resinoids are sufficiently volatile, they could be steam distilled into an essential oil (Anonymous, 2004a).

Concretes are liquid, semiliquid, or solid materials that contain odorless fats, waxes, color pigments, and essential oil. In the case of *Jasmine grandiflorum,* concretes contain as much as 55 percent of the volatile oil (McMahon, 1998). As the concretes are not widely used in perfumery, they have to be extracted with a polar solvent such as alcohol to produce absolute (Lawrence, 1995). As the fatty acids and waxes are not alcohol soluble materials, they should be removed. In general, the waxy concrete matter is warmed and agitated with pure ethanol. Different temperature ranges, 30 to 60°C (Lawrence, 1995) and 46 to 52°C for jasmine concrete (McMahon, 1998), are reported for warming up the concrete materials. The obtained semiliquid or liquid solution is then cooled to between −5 to −35.5°C, while the solution is kept agitated (Lawrence, 1995; McMahon, 1998). Most of the waxes are cold filtered leaving miscella consisting of absolute and alcohol. The alcohol is removed by distillation at low temperature and reduced pressure. The absolutes containing a small amount of nonvolatile matter such as waxes are produced after removing the alcohol through an evaporation process (Davis et al., 2003; Anonymous, 2004a). The solvent recovered from the gentle distillation process is pumped to the solvent tank for further use.

In general, absolutes are highly concentrated viscous liquids. About 100 kg of fresh rose flowers and 150 kg of rose concretes are needed to produce 1 kg of rose concrete and rose absolutes, respectively (Anonymous, 2004b). Another interesting observation is one reported by McMahon (1998) stating that approximately 8,000,000 hand-harvested jasmine flowers are needed to produce 1 kg of jasmine absolute.

An absolute is not a pure essential oil. Depending on the source plant material, actual volatile oil content of the absolutes ranges from 10 to 55 percent (McMahon, 1998). Typically they contain up to 5 percent polar solvents like ethanol, some waxes and sometimes, trace of nonpolar solvents like hexane, depending on the source plant material and the method of extraction used (Anonymous, 2001b; Anonymous, 2004a). Examples of plants that are processed into absolutes are jasmine, rose, and orange blossoms. The diagrammatic representation of the main processing steps of the concrete and absolute extraction processes are given in Figure 5.21.

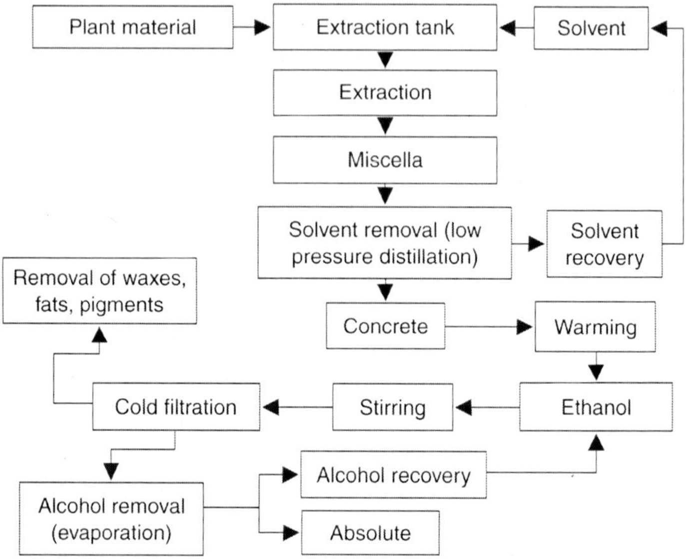

FIGURE 5.21. The simplified flow diagram of the concrete and absolute extraction processes. *Source:* Adapted from Lawrence, 1995.

Extraction with Cold Fat-Enfleurage

Enfleurage extraction uses vegetable and animal cold fats to extract fragrance materials from blossoms. This is a very labor-intensive and expensive extraction technique compared to other extraction methods and used only to extract certain blossoms like jasmine and roses that have very delicate fragrance components. If they are subjected to distillation or solvent extraction the valuable and too delicate aromatic components of these blossoms will be destroyed (McMahon, 1998). In this extraction process, freshly harvested flowers are placed on trays containing fats or lard (Anonymous, 2004c). The trays are then stacked on top of each other to keep out the air and stored in wooden frames (Debbie, 2000; Anonymous, 2004c). The flowers remain on this greasy compound for several duration times ranging from 1 to 60 days (Anonymous, 2004d). Depending on the flower type, every morning or several days later (12-30 h for jasmine, 24-100 h for tuberose) extracted flowers are replaced with freshly harvested blossoms (Sozio, 1956; Anonymous, 2003b). This process is repeated until the base material is saturated with the essential oil. It can take 24 to 36 batches before the base material is saturated (Lawrence, 1995). Saturated base materials are scraped from the trays and the resulting products called pomade are then melted under very low heat, filtered, and further extracted with alcohol to remove unwanted content like pigments, fats, and waxes (Anonymous, 2004b). In the final step the alcohol is evaporated through a distillation process at low temperatures and reduced pressure (Debbie, 2000; Anonymous, 2003b). The resulting concentrated product is called absolute. Approximately 1 ton of freshly harvested jasmine blossoms are needed to produce just 1 liter of absolute (Anonymous, 2004c). Examples of flowers processed by enfleurage extraction are jasmine, violet, tuberose, and rose. However, the enfleurage extraction method is almost obsolete today and has been replaced by inexpensive and more efficient modern solvent extraction techniques.

Extraction with Hot or Warm Fluids/Maceration

Maceration can be described as a process of extracting fragrant oils from plant material by soaking in warm or hot fluids and agitating the plant material and the solvent together. It is very similar to the cold fat extraction method. In this process usually hot or warm oil is used. The process is very time consuming. After completing the maceration the solution is cooled and filtered to separate the fragrant oil. The resulting product called pomade or infusion is then further processed to separate the essential oil. In the case of maceration with hot or warm water, at the end of the process, the solvent is drained off and the miscella that is obtained is subjected to centrifugation and expression. This technique is no longer in use today and has been replaced by modern solvent extraction (Lawrence, 1995).

Ultrasonically Assisted Solvent Extraction

Use of ultrasound in the extraction process, also known as sonication, is not a new technology. It has been used since the 1950s, but is not yet in common usage (Vinatoru et al., 1997; Salisova et al., 1997; Valachovic et al., 2001; Vinatoru, 2001). Current studies have focused on the use of ultrasound to enhance the solvent extraction process. The possible extraction processes for ultrasonic enhancement are extraction with light solvents such as petroleum ether, extraction with water, and extraction with water-alcohol (Vinatoru et al., 1997).

The mechanism of the ultrasonic extraction has been discussed in detail by several authors (Vinatoru et al., 1997; Luque-Garcia and Luque de Castro, 2003). When the essential oil containing plant material is subjected to ultrasound, external essential oil glands can easily be destroyed by sonication and allow the rinsing of the cell content. On the other hand, in the case of essential oil containing internal glands, the milling degree of the plant material determines the extraction level. By milling the plant material, the surface area exposed to ultrasound as well as solvent will increase (Vinatoru et al., 1997). The most possible mechanism of ultrasound extraction suggests that the ultrasound application intensifies the mass transfer and improves the penetration of the solvent into the cell material and reduces the time for the diffusion process which yields grease (Vinatoru et al., 1997, 1999).

Ultrasonically assisted extraction uses bath or probe devices for the application of ultrasound. In general, frequency ranges from 20 to 500 kHz are used for ultrasonically assisted extraction processes (Vinatoru et al., 1997; Salisova et al., 1997; Valachovic et al., 2001; Vinatoru, 2001; Toma et al., 2001). Sonication time during ultrasonically assisted extraction depends on the plant material to be extracted. Although continuous sonification provides better yields, due to economical concerns, short sonification times such as 30 min are preferred in industrial ultrasound-assisted extraction systems (Valachovic et al., 2001; Vinatoru, 2001). This technique could reduce the extraction time—in some cases up to 32 times shorter extraction time—and increase the product yields significantly (Vinatoru, 2001), but needs to be further studied to become fully commercialized.

NEW EXTRACTION TECHNIQUES

Supercritical Fluid Extraction

There is an increasing interest in supercritical fluid extraction of MAP in recent years. Supercritical fluid extraction uses supercritical fluids to extract the useful components of the plant materials and provides more selective, efficient, and faster extraction compared to conventional solvent extraction methods. As shown in Figure 5.22, the supercritical phase is defined by the supercritical temperature and supercritical pressure. Any substance reaches the supercritical phase when its temperature and the pressure are raised above its critical values (Table 5.1). Ideally, a supercritical phase can be termed as an intermediate state between gas and liquid (Fukushima, 1999). At that phase, the solvating power of supercritical fluids is liquidlike, while its transport properties (diffusivity and viscosity) are gaslike (Bevan and Marshall, 1994; Sihvonen et al., 1999; Chen and Ling, 2000; Lang and Wai, 2001).

Supercritical CO_2 Extraction

Carbon dioxide is the most widely used supercritical fluid for the extraction of MAP because of its easily achievable low critical parameters, low cost, and so on. Compared to other chemical solvents, supercritical CO_2 among the other supercritical fluids offers many advantages for the extraction of natural fragrant compounds from MAP. The advantages of supercritical CO_2 may be summarized as follows:

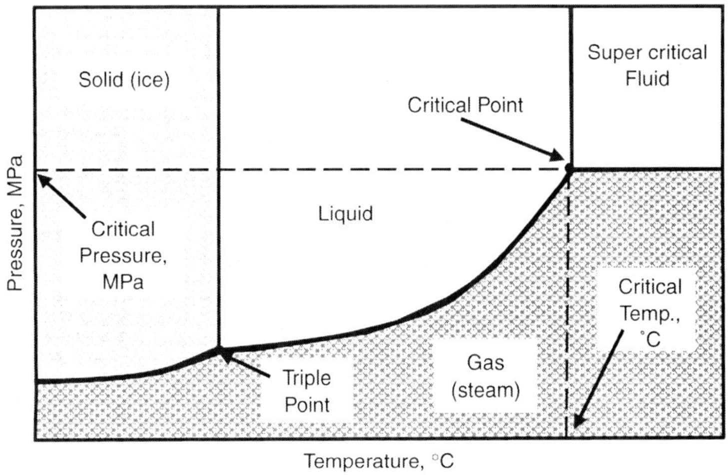

FIGURE 5.22. Simplified pressure/temperature diagram for carbon dioxide.

TABLE 5.1. The critical temperature and the pressure values for some common supercritical fluids (Fukushima, 1999; Williams and Sprenger, 1999; Ghaderi, 2000; Demirbas, 2001; Hu et al., 2003).

Solvent	Critical temperature, °C	Critical pressure, MPa
Ammonia	132.5	11.399
Carbon dioxide	31.3	7.387
Ethane	32.3	4.874
Ethylene	9.2	5.034
Propane	96.7	4.246
Pentane	196.6	3.374
Water	374.2	22.099

- highly adjustable solvent power/selectivity; allows the extraction of a wide range of complex compounds including plant materials (Fukushima, 1999; Sihvonen et al., 1999; Bevan and Marshall, 1994; Phelps et al., 1996; Reverchon et al., 1993)
- easily recyclable and recoverable from the extract leaving no harmful solvent residue, minimizes the waste generation (Stashenko et al., 1996; Sihvonen et al., 1999; Fukushima, 1999)
- odorless, colorless, nontoxic, nonflammable, and nonexplosive (Goto et al., 1997; Sihvonen et al., 1999; Chen and Ling, 2000; Wong et al., 2001)
- higher diffusion coefficient, lower viscosity, surface tension, and molecular weight compared to liquid solvents; leads liquid-like salvation power, gaslike diffusivity and viscosity (Bevan and Marshall, 1994; Sihvonen et al., 1999; Chen and Ling, 2000; Lang and Wai, 2001)
- ecologically friendly process, does not increase CO_2 emissions (Wong et al., 2001; Lang and Wai, 2001)
- readily available and inexpensive (Goto et al., 1997; Sihvonen et al., 1999)
- allows simultaneous fractionation and deterpenation of extracts (Sato et al., 1998; Reverchon et al., 1999; Luque de Castro et al., 1999)
- natural and regarded as safe gas for natural products (Sanders, 1993)

- usually performed at low temperatures, allows the low temperature extraction of thermally labile compounds with minimal damage (Goto et al., 1997; Sihvonen et al., 1999; Chen and Ling, 2000; Marr and Gamse, 2000; Wong et al., 2001)
- allows the extraction of high boiling point components at relatively low temperatures (Fukushima, 1999; Lang and Wai, 2001)
- produces better quality extracts with a delicacy and freshness that is close to the natural source material (Anonymous, 2003d)
- provides highly flexible automated processing conditions (Wong et al., 2001)
- highly reduces the processing time (Sihvonen et al., 1999; Chen and Ling, 2000)
- nonpolar; when coupled with the other chemical solvents, solubility of polar components and the selectivity of the process can be increased (Vannoort et al., 1990; Bevan and Marshall, 1994; Phelps et al., 1996; Sihvonen et al., 1999)
- required working conditions are easily achievable (Sihvonen et al., 1999; Wong et al., 2001)
- longer shelf life (Reverchon, 1997)
- suitable for direct coupling with a chromatograph allowing immediate analysis after extraction (Vannoort et al., 1990; Bevan and Marshall, 1994; Phelps et al., 1996; Lang and Wai, 2001; Wong et al., 2001)
- easy optimization of extraction process (Lang and Wai, 2001)

Supercritical carbon dioxide extraction can be carried out both as a batch or continuous process. Extraction time is generally short (15 to 120 min or several hours) and depends highly on the extraction conditions, type of operating mode (batch or continuous), extraction solvent, co-solvent, and the like (Reverchon, 1997; Maheshwary, 1998; Chen and Ling, 2000). Most of the supercritical CO_2 applications extract the useful plant components at pressures ranging from 100 to 500 bars and the temperatures are between 40 and 80°C (Reverchon, 1997; Maheshwary, 1998; Smith, 1999; Chen and Ling, 2000; Lang and Wai, 2001; Vasukumar and Bansal, 2003). For the extraction of essential oils by supercritical CO_2, the suggested density of the solvent ranges from 0.25 to 0.50 g/cm^3 (Reverchon, 1997).

A simplified diagrammatic representation of the supercritical CO_2 extraction process is shown in Figure 5.23. In this process, the plant material is filled into the high pressure stainless steel extraction tank, and then the supercritical carbon dioxide is pumped through a heat exchanger into the tank at a certain temperature and pressure (above the critical point; see Table 5.1). The temperature and the pressure variations inside the extraction vessel should be carefully controlled during the extraction process. The aromatic flavoring constituents of the plant material are then dissolved in supercritical carbon dioxide. The resulting extract rich in CO_2 is transferred to a separation tank that is maintained below the critical point. The CO_2 is then removed from the extract by means of a temperature/pressure change. The recovered CO_2 is transferred to the reservoir for reuse.

Supercritical CO_2 extraction of plant material is now becoming a commercially viable process because of the recent developments and advances in process technology and the legal limitations of solvent residues in processed plant materials. Tea and coffee decaffeination and hop extraction are some well-known examples of fully developed commercial supercritical fluid extraction technologies (Phelps et al., 1996; Marr and Gamse, 2000; Sihvonen et al., 1999). Other known commercial applications are supercritical fluid extraction of oleoresins and essential oils (Reverchon, 1997; Marr and Gamse, 2000; Imison and Unthank, 2000). However, even though a highly increasing tendency is observed in patent applications in this area (Fukushima, 1999; Sihvonen et al., 1999), commercial application of supercritical fluid extraction is highly restricted by the high capital investment, high pressure equipment, requiring sensitive process control, and

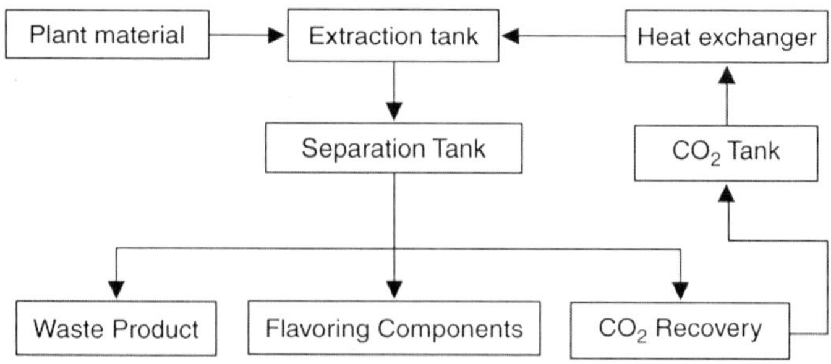

FIGURE 5.23. Schematic representation of supercritical CO_2 extraction process.

maintenance and safety problems in design and operation. Therefore, supercritical fluid extraction of essential oils in industrial scale is not yet a fully developed, mature technology. A feasibility analysis showed that the extraction of various plants such as cloves, cumin, cardamom, and ginger by using a 3 × 500 liters capacity extractor has around 2.2 million US$ machinery/equipment cost with the annual operational cost ranging from 4 to 6 million US$ and the payback period ranging from 2.5 to 3.5 years (Anonymous, 2004e).

It should be noted that a considerable amount of research is needed to obtain more reliable and extensive data on the solubility of essential oils to scale-up the extraction process (Reverchon, 1997; Vasukumar and Bansal, 2003). There are a number of publications on the extraction of essential oils, flavors, and other useful components from various MAP in an analytical scale. Comprehensive overviews of supercritical fluid (CO_2) extraction of MAP have been given by several researchers (Reverchon, 1997; Smith, 1999; Luque de Castro et al., 1999; Chen and Ling, 2000; Marr and Gamse, 2000; Lang and Wai, 2001). These plants include chamomile flowers, angelica roots, lavandin, lavender flower, lavandula flower, thyme leaves, rosemary leaves, rose flowers, sage leaves, savory leaves, sandalwood, spearmint leaves, clove, peppermint, lemon, citrus, basil leaves, cardamom seeds, cinnamon leaf, geranium leaves, jasmine flowers, bergamot, eucalyptus, primrose, valerian, dandelion, coriander fruit, dragonhead leaves, yarrow, St. Mary's thistle, anise seeds, juniper, paprika, marjoram leaves, oregano leaves, onion bulbs, paprika fruit, pepper fruit, yew tree needles, ginger rhizomes, caraway seeds, iris rhizomes, hops fruit, and so on.

The performance of the supercritical fluid extraction system depends to a great extent on several process parameters. These factors can be summarized as follows (Reverchon, 1997; Smith, 1999; Chen and Ling, 2000; Lang and Wai, 2001; Povh et al., 2001; Vasukumar and Bansal, 2003):

- type of extraction solvent; choice of main extraction solvent, type and amount of modifier or co-solvent
- extraction conditions; temperature, pressure, extraction time, solvent flow rate
- particle size and distribution
- system leaks and system contamination

The supercritical CO_2 extraction produces total extracts like concrete if only a single stage separation is used and the residence time of the supercritical solvent is sufficiently long (Goto et al., 1997; Reverchon, 1997; Smith, 1999; Povh et al., 2001). Therefore, this product should be further processed into essential oil by fractional separation with the extraction system having two or more separators in series (Reverchon, 1997). In general, increasing the temperature and the pressure provides better extraction efficiency. Nevertheless, higher extraction temperatures should be avoided because of the thermally sensitive essential oil compounds. Similarly, safety concerns also limit the usage of higher extraction pressures (Chen and Ling, 2000). Increasing the CO_2 density may lead to quality degradation due to the co-extraction of some unwanted compounds like waxes and triglycerides (Bartley and Foley, 1994; Luque de Castro et al., 1999). Based on the extensive experimental results on supercritical fluid extraction, optimal conditions for essential oil extraction are at pressures lower than 100 bar (80 to 100 bar) and at temperatures between 40 and 50°C (Reverchon et al., 1993; Reverchon, 1997).

Supercritical carbon dioxide has non polar solvent properties and also has the ability to dissolve some low-polarity compounds such as waxes, fatty acids, color pigments, and resins in minor quantities (Luque de Castro et al., 1999; Lang and Wai, 2001). It is possible to control the content of these unwanted compounds by using suitable extraction temperatures and pressures, multistage separation, cosolvents, and adsorbent materials like silica gel. These will be discussed below.

Solvating power or selectivity of the supercritical fluid can be increased or modified by using a small amount of co-solvent or modifier (3-20 percent) (Reverchon, 1997; Fukushima, 1999; Chen and Ling, 2000; Luque de Castro and Jimenez-Carmona, 2000; Lang and Wai, 2001). The processing time can also be decreased by using a modifier solvent, due to the increased solvent selectivity (Sihvonen et al., 1999). Methanol, acetone, acetonitrile, freon-22, nitrous oxide, and water are some examples of modifier solvents. Among them methanol is the most widely used co-solvent (Reverchon, 1997; Chen and Ling, 2000; Lang and Wai, 2001). On the other hand, modifier solvent usage complicates the extraction system (a further process step is needed for their elimination) and increases the equipment and operational costs (Reverchon, 1997; Sihvonen et al., 1999; Luque de Castro and Jimenez-Carmona, 2000).

Another alternative method to modify or improve the selectivity of the supercritical fluids is utilization of adsorbents such as silica gel, aluminum oxide, Kieselgur, cellulose, bentonite, magnesium silicate, magnesium sulphate, and calcium sulphate in the supercritical extraction process (Bart et al., 1994; Chouchi et al., 1995, 1996; Dugo et al., 1995; Sato et al., 1998). It is reported that silica gel is used in citrus oil processing to extract the oxygenated aroma compounds and yield high quality essential oil with less terpenes and less nonvolatile components (Bart et al., 1994; Chouchi et al., 1995, 1996; Sato et al., 1998). It is further stated that by using silica gel in supercritical fluid extraction, compared to terpenes, more selective extraction of aroma compounds could be possible at relatively higher pressures which is required to regenerate the adsorbent and maintain the activity of the adsorbent (Sato et al., 1998).

Subcritical Water Extraction

Subcritical or super heated water extraction uses water at elevated temperatures between 100 and 374°C (see Table 5.1) and at certain pressures high enough to maintain the liquid state (Gamiz-Gracia and Luque de Castro, 2000). Under these conditions water can be used as a low polarity extraction solvent (Smith, 2002).

Subcritical water for the extraction of essential oils from plant material is reported as a promising alternative to the conventional extraction techniques, as well as supercritical CO_2 extraction, because of the following advantages (Basile et al., 1998; Luque de Castro et al., 1999; Rovio et al., 1999; Gamiz-Gracia and Luque de Castro, 2000; Ayala and Luque de Castro, 2001; Kubatova, Lagadec et al., 2001; Kubatova, Miller et al., 2001; Lang and Wai, 2001; Smith, 2002; Ozel, et al., 2003):

- higher extraction ability for polar compounds
- avoids the extraction of cuticular waxes and lipids in a single stage extraction
- much faster, cheaper and cleaner (solvent free) than other methods
- environmentally friendly
- nontoxic

- suitable for automation
- requires relatively low pressures compared to supercritical fluid extraction systems
- equipment cost is cheaper than supercritical fluid extraction systems
- provides energy economy
- more efficient extraction method in terms of qualitative composition of the extract containing higher oxygenated compounds and lower terpene fraction; better representation of natural aroma of the essential oil
- eliminates the need for drying stage
- provides selective extraction by changing temperature
- uses relatively low temperatures and avoids losses and degradation of volatile and thermally sensitive compounds

Although subcritical water extraction has several advantages over the conventional and other new extraction techniques, the temperature range used to extract the essential oils is relatively high. The essential oil extraction with subcritical water is usually performed at temperature ranges from 100 to 150°C and at pressures up to 6 MPa (Basile et al., 1998; Luque de Castro et al., 1999; Rovio et al., 1999; Gamiz-Gracia and Luque de Castro, 2000, 2001; Ayala and Kubatova et al., 2001a; Lang and Wai, 2001; Smith, 2002; Ozel et al., 2003). Therefore, such elevated temperatures may not be suitable for the extraction of thermally sensitive essential oils from medicinal and aromatic plants. Ammann et al. (1999) reported that sabine hydrate in peppermint essential oil was lost compared to supercritical fluid extracted and steam distilled peppermint essential oils. Furthermore, compared to steam distillation and supercritical fluid extraction, substantial degradation of linalool and γ-terpinene were reported at temperatures above 150°C for subcritical water extraction of savory and peppermint (Kubatova, Lagadec et al., 2001). Jimenez-Carmona et al. (1999) reported that compared to hydrodistilled marjoram essential oil, the monoterpene hydrocarbons (α-pinene, β-pinene, and β-myrcene) are found in lesser amount in the continuous subcritical water extracted marjoram essential oil, while the oxygenated compounds found in the continuous subcritical water extracted marjoram essential oil are more concentrated. Moreover, Lang and Wai (2001) concluded that this method was not suitable for thermally liable compounds.

Several researchers reported that the reactive nature of the water under subcritical conditions could be corrosive and damage the extraction vessel (Basile et al., 1998; Luque de Castro et al., 1999; Smith, 2002). However, Luque de Castro et al. (1999) state that this drawback of subcritical water extraction can be easily prevented by using ultra pure and degasified water. Naturally, such purifying processes will need additional equipment and increase the equipment and operational cost of the system.

To conclude, subcritical water extraction method has significant advantages over existing extraction techniques, but also exhibits some important limitations that should be solved with the help of new developments. Comprehensive studies should be performed on the extraction of essential oils and other useful compounds from medicinal and aromatic plants.

Microwave-Assisted Extraction

The use of microwave energy for the extraction of organic compounds from MAP is a relatively new technique. The microwave extraction method is quite different from the conventional means and is used for the fastening of the extraction process. Microwaves penetrate directly into the material and lead to internal heat generation by molecular friction resulting from dipolar rotation in polar solvents and from the convective migration of dissolved ions (Franca and Haghighi, 1996). It provides fast and selective heating throughout the material without the need to overheat the atmosphere. In this process, fresh or dried plant material is immersed in a solvent that is transparent to microwaves, and this mixture is then pumped into the process tank where it is subjected to microwave irradiation (Belanger, 2002). When the essential oil containing plant material is subjected to microwave irradiation, the microwave energy is selectively absorbed by the liquid portion of the plant material (essential oils, water molecules, etc.). The temperature of the liquid existing in the plant material increases rapidly and ruptures the essential oil containing gland walls allowing the migration of the essential oils out of the plant matrix into the solvent (Pare et al., 1994). The resulting extract is then filtered and further processed into essential oil. The technique can be coupled with different solvent extraction systems (Luque de Castro et al., 1999). Microwave-assisted extraction provides the following advantages (Camel, 2000; Ayuso-Garcia and Luque de Castro, 2001; Belanger, 2002; Alfaro et al., 2003; Stashenko et al., 2004a,b):

- reduces the processing time and energy consumption
- improves the yield and quality in many cases
- reduces solvent consumption
- reduces investment cost
- easy process optimization
- allows the extraction of both fresh and dried materials
- applicable technology both in analytical and in industrial scale (several units are in commercial operation)

Phytol Extraction

In the late 1980s, the phytol or florasol extraction technique was developed by Dr. Peter Wilde (Correl, 2003). This extraordinary new extraction technique uses a benign gaseous hydroflorocarbon refrigerant solvent (R134a-1,1,1,2-tetrafluoroethane) called florasol with a boiling point of –26.2°C for the extraction of aromatic essential oils and bio-active compounds from MAP materials (Baser, 1999; Correl, 2003; Anonymous, 2004f,g,h,i). This technique is approved for use in food flavoring by the European Commission (Anonymous, 2004i). This new extractant is a clear, nontoxic, odorless, colorless, nonflammable, nonexplosive, ozone friendly, and highly selective inert liquid. In this process, extracted plant material is filled in a sealed extraction vessel, and the extraction solution at a pressure of 5 bars is then injected into the extraction vessel. The resulting extract is transferred to another container where the extraction solvent is evaporated below freezing temperature. The recovered solvent is collected in another chamber for reuse. The extract obtained is a pure florasol, free from waxes or fats, and contains the most delicate clear aromatic essential oil. The extraction process requires low energy and labor cost and is performed at or below ambient temperatures. Thus, there is no thermal degradation of the products (Anonymous, 2004i).

REFERENCES

Alfaro MJ, Belanger JMR, Padilla FC, Pare JRJ. 2003. Influence of solvent, matrix dielectric properties, and applied power on the liquid-phase microwave-assisted processes (MAP™) extraction of ginger (Zingiber officinale). *Food Research International;* 36(5): 499-504.

Ammann A, Hinz DC, Addleman RS, Wai CM, Wenclawiak BW. 1999. Hot water extraction, steam distillation and SFE of peppermint oil. *Fresenius Journal of Analytical Chemistry;* 364: 650-653.

[Anonymous] 2001a. Essential Oils. Available from: http://inspie3.home.mind spring.com/oils.htm. Accessed 2003 June 11.

[Anonymous] 2001b. Absolutes. Available from: http://essentialoils.org/absolutes. htm. Updated; Accessed 2004 February 17.

[Anonymous] 2003a. Purveyors of Organic and Wild-Crafted Essential Oils and Healing Accessories. Available from: http://www.aromawerks.com/Distillation .htm. Accessed 2003 June 02.

[Anonymous] 2003b. Medicinal HerbFAQ Part 5/7. Available from: www.cs.uu .nl/wais/html/na-dir/medicinal-herbs/part5.html. Accessed 2003 October 05.

[Anonymous] 2003c. The Extraction of Essential Oils. Available from: http:// www.oneplanetnatural.com/stores/oneplanet/distillation.htm. Accessed 2003 March 27.

[Anonymous] 2003d. Topics in Food Chemistry. Available from: http://alfa.ist .utl.pt/~fidel/flaves/sec5/sec5.html. Accessed 2003 21 May.

[Anonymous] 2003e. How are essential oils made. Available from: http://www .deancoleman.com/whatareoils.htm. Accessed 2003 June 16.

[Anonymous] 2004a. Amateur Aromatherapy. Available from: http://www.andy barson.btinternet.co.uk/intro.htm. Accessed 2004 April 08.

[Anonymous] 2004b. Essential Oil Directory. Available from: http://www.frontier coop.com/ac/eo.html. Accessed 2004 May 28.

[Anonymous] 2004c. L'enfleurage. Available from: http://www.museesdegrasse .com/MIP/fla_ang/techno_enfleurage.shtml. Accessed 2004 June 13.

[Anonymous] 2004d. Leading techniques; Enfleurage. Available from: http://www .ac-orleans-tours.fr/physique/docly/divers/parfum/enfleurage.html. Accessed 2004 June 15.

[Anonymous] 2004e. Supercritical Fluid Extraction (SCFE) Technology for Natural Product Extracts. Asian and Pacific Centre for Transfer of Technology. Technology Offer Ref.ID: APC-5040-TO. Available from: http://www.apctt.org/ database.html. Accessed 2004 February 17.

[Anonymous] 2004f. Wilde & Company's Florasol. Available from: http://www .essentialoil.com. Accessed 2004 March 23.

[Anonymous] 2004g. Balancing Body, Mind & Spirit With Pure Essential Oils. Methods of Extracting Essential Oils. Available from: http://www.naturegift .com. Accessed 2004 May 21.

[Anonymous] 2004h. Leydet Aromatics, Aromatherapy. Available from: http:// www.leydet.com. Accessed 2004 May 11.

[Anonymous] 2004i. Aromatic Oil Extractor. Available from: http://www.chemplant.co.uk/cp-prod1.htm. Accessed 2004 July 08.

Ayala RS, Luque de Castro MD. 2001. Continuous subcritical water extraction as a useful tool for isolation of edible essential oils. *Food Chemistry;* 75(1): 109-113.

Ayuso-Garcia LE, Luque de Castro MD. 2001. Employing focused microwaves to counteract conventional Soxhlet extraction drawbacks. *Trends in Analytical Chemistry;* 20(1): 28-34.

Bart D, Chouchi D, Porta GD, Reverchon E, Perrut MC. 1994. Desorption of lemon peel by supercritical carbon dioxide: Deterpenation and psoralens elimination. *Journal of Supercritical Fluids;* 7(3): 177.

Bartley JP, Foley P. 1994. Supercritical fluid extraction of Australian-grown ginger. *Journal of the Science of Food and Agriculture;* 66: 365.

Başer KHC, Kürkçüoğlu M, Konur OZ. 1990. Türk Gül Yağinin Üretimi ve Özellikleri. Anadolu Üniversitesi Tıbbi Bitkiler Araştırma Merkezi. Tıbbi ve Aromatik Bitkiler Bülteni, Gül Özel Sayısı, Temmuz; 4: 13-15.

Baser KHC. 1999. Industrial Utilization of Medicinal and Aromatic Plants. Proceedings of Second World Congress on Medicinal and Aromatic Plants for Human Welfare, 10-15 November 1997, Mendoza, Argentina. Martino V, Bandoni A, Blaak G, Capelle N, editors. *Acta Hort.* 503: 21-29.

Basile A, Jimenez-Carmona MM, Clifford AA. 1998. Extraction of rosemary by superheated water. *Journal of Agriculture and Food Chemistry;* 46(12): 5205-5209.

Belanger JMR. 2002. MAP™ Microwave Assisted Process. Available from: http://www.oceta.on.ca/profiles/cwt-tran/map.html. Accessed 2002 June 19.

Bevan CD, Marshall PS. 1994. The use of supercritical fluids in the isolation of natural products. *Natural Product Reports;* 32: 451-466.

Camel V. 2000. Microwave-assisted solvent extraction of environmental samples. *Trends in Analytical Chemistry;* 19(4): 229-248.

Chen YT, Ling YC. 2000. An overview of supercritical extraction in Chinese herbal medicine: from preparation to analysis. *Journal of Food and Drug Analysis;* 8(4): 235-247.

Chouchi D, Bart D, Reverchon E, Porta GD. 1995. Supercritical CO_2 desorption of bergamot peel oil. *Industrial Engineering Chemistry Research;* 34(12): 4508-4513.

Chouchi D, Bart D, Reverchon E, Porta GD. 1996. Bigarade peel oil fractionation by supercritical CO_2 desorption. *Journal of Agricultural and Food Chemistry;* 44: 1100-1104.

Chung B. 2000. Natural Plant Extracts: Export Market Opportunities in the USA. A Report for the Rural Industries Research and Development Corporation, July 2000, RIRDC Publication No: 00/51, RIRD Project No: BRA-3A, P. 49. Available from: http://www.rirdc.gov.au/reports/Index.htm. Accessed 2003 April 14.

Correl A. 2003. Aromatheraphy and practical uses in today's World. Available from: http://www.phancypages.com/newsletter/98.ArleneAromaTheraphy.htm. Accessed 2003 June 11.

Dachler M, Pelzmann H. 1989. Heil -und Gewürzpflanzen, Anbau-Ernte–Aufbereitung: Österreichischer Agrarverlag, Wien; 40p.

Davis E, Hassler J, Ho P, Hover A, Kruger W. 2003. Essential Oils. Available from: http: //www.wsu.edu/~gmhyde/433_web_pages/433Oil-web-pages/essence/essence-oils.html. Accessed 2003 July 23.

Debbie P. 2000. Scents-ability: all you wanted to know about fragrance, and more, Part 1. *Cosmetics, Don Mills;* 28(3): 16.

Demirbas A. 2001. Supercritical fluid extraction and chemicals from biomass with supercritical fluids. *Energy Conversion & Management;* 42: 279-294.

Dugo P, Mondello L, Bartle KD, Clifford AA, Breen DGPA, Dugo G. 1995. Deterpenation of sweet orange and lemon essential oils with supercritical carbon dioxide using silica gel as an adsorbent. *Flavour and Fragrance Journal;* 10: 51-58.

FAO. 1992. Minor Oil Crops, Essential Oils, Part III. Minor Oil Crops, Part I–Edible Oils, Part II–Non Edible Oils, Part III–Essential Oils. *FAO Agricultural Services Bullettin;* 94: 175-2417.

Franca AS, Haghighi K. 1996. Adaptive finite element analysis of microwave driven convection. *International Communications in Heat and Mass Transfer;* 23(2): 177-186.

Fraser S, Whish JPM. 1997. A Commercial Herb Industry for NSW–an Infant Enterprise. A Report for the Rural Industries Research and Development Corporation, May 1997, RIRDC Research Paper Series No: 97/18, 141p. Available from: http://www.rirdc.gov.au/reports/NPP/UNE-30A.DOC. Accessed 2003 June 24.

Fukushima Y. 1999. Application of Supercritical Fluids–Review. *R&D Review of Toyota CRDL;* 35(1): 1-9.

Gamiz-Gracia L, Luque de Castro MD. 2000. Continuous subcritical water extraction of medicinal plant essential oil: comparison with conventional techniques. *Talanta;* 51(6): 1179-1185.

Ghaderi R. 2000. A supercritical fluids extraction process for the production of drug loaded biodegradable microparticles. Acta Universitatis Upsaliensis. *Comprehensive Summaries of Uppsala Dissertations from the Faculty of Pharmacy.* 234, 46p. Uppsala. ISBN 91-554-4800-3.

Goto M, Sato M, Kodama, A, Hirose T. 1997. Application of supercritical fluid technology to citrus oil processing. *Physica B* 239: 167-170.

Hu ZQ, Yang JC, Li YG. 2003. Crossover SAFT equation of state for pure supercritical fluids. *Fluid Phase Equilibria;* 205: 1-15.

Imison B Unthank D. 2000. Adding Value to Essential Oils & Other Natural Ingredients. A Report for the Rural Industries Research and Development Corporation, April 2000, RIRDC Publication No: 00/40, RIRD Project No: DAV-97A, 46p. Available from: http://www.rirdc.gov.au/reports/Index.htm. Accessed 2002 June 21.

Jimenez-Carmona MM, Ubera JL, Luque de Castro MD. 1999. Comparison of continuous subcritical water extraction and hydrodistillation of marjoram essential oil. *Journal of Chromatography A;* 855(2): 625-632.

Kubatova A, Lagadec AJM, Miller DJ, Hawthorne SB. 2001. Selective extraction of oxygenates from savory and peppermint using subcritical water. *Flavour and Fragrance Journal;* 16(1): 64-73.

Kubatova A, Miller DJ, Hawthorne SB. 2001. Comparison of subcritical water and organic solvents for extracting kava lactones from kava root. *Journal of Chromatography A;* 923 (1-2): 187-194.

Kunkar A, Başer KHC, Tanrıverdi H. 1993. Narenciye Ürünleri Teknolojileri-I. Anadolu Üniversitesi Tıbbi ve Aromatik Bitkiler Araştırma Merkezi, Tıbbi ve Aromatik Bitkiler Bülteni, Ocak'93; *Sayı*: 7-8: 19-26.

Lang Q, Wai CM. 2001. Supercritical fluid extraction in herbal and natural product studies–a practical review. *Talanta;* 53(4): 771-782.

Lawrence BM. 1995. The isolation of aromatic materials from natural plant products. In: *A Manual on the Essential Oil Industry.* K Tuley De Silva, Editor. UNIDO, Vienna, Austria. 57-154.

Luque de Castro MD, Carmona MMJ, Perez VF. 1999. Towards more rational techniques for the isolation of valuable essential oils from plants. *Trends in Analytical Chemistry;* 18(11): 708-716.

Luque de Castro MD, Jimenez-Carmona MM. 2000. Where is supercritical fluid extraction going? *Trends in Analytical Chemistry;* 19(4): 223-228.

Luque-Garcia JL, Luque de Castro MD. 2003. Ultrasound: a powerful tool for leaching. *Trends in Analytical Chemistry;* 22(1): 41-47.

Maheshwary RC. 1998. Profiles of Patented Technologies of IIT Delhi. Title: Supercritical Fluid Extraction (SCFE) Technology for Agro-Food Product's Extracts. Available from: http://www.fitt-iitd.org/patentedTech/patentTech.htm. Accessed 2003 October 23.

Marr R, Gamse T. 2000. Use of supercritical fluids for different processes including new developments–a review. *Chemical Engineering and Processing;* 39(1): 19-28.

McMahon C. 1998. Aromatic Absolutes: Their Role in Natural Perfumery. Available from: http://members.aol.com/parijata/absolutes.html. Accessed 2000 July 19.

Ozel MZ, Gogus F, Lewis AC. 2003. Subcritical water extraction of essential oils from Thymbra spicata. *Food Chemistry;* 82(3): 381-386.

Pare JRJ, Belanger JMR, Stafford SS. 1994. Microwave assisted process (MAP™): a new tool for the analytical laboratory. *TrAC Trends in Analytical Chemistry;* 13(4): 176-184.

Phelps CL, Smart NG, Wai CM. 1996. Past, present and possible future applications of supercritical fluid extraction technology. *Journal of Chemical Education;* 73(12): 1163-1168.

Povh NP, Marques MOM, Meireles, MAA. 2001. Supercritical CO_2 extraction of essential oil and oleoresin from chamomile (*Chamomilla recutita* [L.] Rauschert). *The Journal of Supercritical Fluids;* 21(3): 245-256.

Reverchon E. 1997. Supercritical fluid extraction and fractionation of essential Oils and related products. *The Journal of Supercritical Fluids;* 10(1): 1-37.

Reverchon E, Della Porta G, Lamberti G. 1999. Modeling of orange flower concrete fractionation by super critical CO_2. *The Journal of Supercritical Fluids;* 14(2): 115-121.

Reverchon E, Donsi G, Osseo LS. 1993. Modeling of supercritical fluid extraction from herbaceous matrices. *Industrial & Engineering Chemistry Research;* 32(12): 2721-2726.

Rinder R, Bomme U. 1998. Wasserdampf - Destillation ätherischer Öle aus frischen oder angewelkten Pflanzen, Merkblatt der Bayerischen Landesanstalt für Bodekultur und Pflanzenbau, Postfach 1641, D-85316 Freising.

Rovio S, Hartonen K, Holm Y, Hiltunen R, Riekkola ML. 1999. Extraction of clove using pressurized hot water. *Flavour and Fragrance Journal;* 14(6): 399-404.

Runham SR. 1996. An updated Review of the Potential Uses of Plants Grown for Extracts Including Essential Oils and Factors Affecting Their Yield and Composition, December, 1996. MAFF Ref ST0105 ADAS Desk Study. 75p. Available from: http://www.defra.gov.uk/farm/acu/research/reports/RDREP03.PDF. Accessed 2003 April 20.

Salisova M, Toma S, Mason TJ. 1997. Comparison of conventional and ultrasonically assisted extractions of pharmaceutically active compounds from *Salvia officinalis. Ultrasonics Sonochemistry;* 4(2): 131-134.

Sanders N. 1993. Food legislation and the scope for increased use of near-critical fluid extraction operations in the food, flavoring and pharmaceutical industries. In *Extraction of Natural Products Using Near-Critical Solvents.* King MB, Bot TR, Editors. Chapman & Hall Inc., New York, USA, 325 pp.

Sato M, Goto M, Kodama A, Hirose T. 1998. New fractionation process for citrus oil by pressure swing adsorption in supercritical carbon dioxide. *Chemical Engineering Science;* 53(24): 4095-4104.

Sihvonen M, Järvenpää E, Hietaniemi V, Huopalahti R. 1999. Advances in supercritical carbon dioxide technologies–review. *Trends in Food Science & Technology;* 10(6-7): 217-222.

Simon JA. 1997. New Crop Introduction: Exploration, Research and Commercialization of Aromatic Plants in the New World. Proceedings of the Prairie Medicinal and Aromatic Plants Conference '97. March 9-11, 1997 at the Royal Oak Inn, Brandon, Manitoba. Available from: http://www.gov.mb.ca/agriculture/crops/alternativecrops/pmap/bkp00s15.html. Accessed 2003 May 21.

Smith RM. 1999. Supercritical fluids in separation science–the dreams, the reality and the future. *Journal of Chromatography A;* 856(1-2): 83-115.

Smith RM. 2002. Extractions with superheated water. *Journal of Chromatography A;* 975(1): 31-46.

Sozio H. 1956. Essential oils. Products of Enfleurage. *Perfume Essential Oil Research;* 47: 160-162. In: Lawrence BM. 1995.

Stashenko EE, Jaramillo BE, Martinez JR. 2004a. Analysis of volatile secondary metabolites of Xylopia aromatica (Lamark) by different extraction and headspace methods and gas chromatography. *Journal of Chromatography A;* 1025(1): 105-113.

Stashenko EE, Jaramillo BE, Martinez JR. 2004b. Comparison of different extraction methods for the analysis of volatile secondary metabolites of Lippia alba (Mill.) N.E. Brown, grown in Colombia, and evaluation of its in vitro antioxidant activity. *Journal of Chromatography A;* 1025(1): 93-103.

Stashenko EE, Puertas MA, Combariza MY. 1996. Volatile Secondary Metabolites from Spilanthes americana Obtained by Simultaneous Steam Distillation–Solvent Extraction and Supercritical Fluid Extraction, *Journal of Chromatography A;* 752(1-2): 223-232.

Toma M, Vinatoru M, Paniwnyk L, Mason TJ. 2001. Investigation of the effects of ultrasound on vegetal tissues during solvent extraction. *Ultrasonics Sonochemistry;* 8(2): 137-142.

Valachovic P, Pechova A, Mason TJ. 2001. Towards the industrial production of medicinal tincture by ultrasound assisted extraction. *Ultrasonics Sonochemistry;* 8(2): 111-117.

Vannoort RW, Chervet JP, Lingeman H, DeJong GJ, Brinkman UA. 1990. Coupling of Supercritical Fluid Extraction with Chromatographic Techniques. *Journal of Chromatography A;* 505(1): 45-77.

Vasukumar K, Bansal AK. 2003. Supercritical fluid technology in pharmaceutical research–Review Article. *CRIPS;* 4 (2): 1-12.

Vinatoru M. 2001. An overview of the ultrasonically assisted extraction of bioactive principles from herbs. *Ultrasonics Sonochemistry;* 8(3): 303-313.

Vinatoru M, Toma M, Mason TJ. 1999. Ultrasonically assisted extraction of bioactive principles from plants and their constituents. *Advances in Sonochemistry,* JAI Press, London; 5.

Vinatoru M, Toma M, Radu O, Filip PI, Lazurca D, Mason TJ. 1997. The use of ultrasound for the extraction of bioactive principles from plant materials. *Ultrasonics Sonochemistry;* 4(2): 135-139.

Vukic R, Perunicic Mi, Brujic B. 1995. Distillation and extraction plants in a single unit. *Plant Report* (In Serbian); 2(2): 39-43.

Williams JM, Sprenger GH. 1999. The 4th state of the matter: The delta state. Physics abstract, physics/9904001. Available from: http://arxiv.org/abs/physics/9904001. Accessed 2003 May 16.

Wong V, Wyllie SG, Cornwell CP, Tronson D. 2001. Supercritical Fluid Extraction (SFE) of monoterpenes from the leaves of *Melaleuca alternifolia* (Tea Tree). *Molecules;* 6: 92-103.

Chapter 6

Industrial Utilization

Yurtsever Soysal
Serdar Öztekin

INTRODUCTION

A broad range of plants are used for medicinal and aromatic purposes, namely herbs, spices, vegetables, trees, natural gums and resins, algae, flowering plants, gymnosperms, ferns, bryophytes, mushrooms, and the like. In general, the whole plant or various plant parts are used for medicinal and aromatic raw material, Figure 6.1.

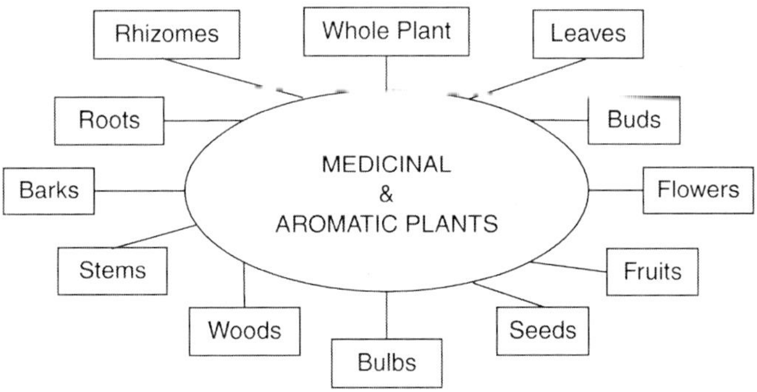

FIGURE 6.1. Most frequently used plant parts as medicinal and aromatic raw material.

Medicinal and Aromatic Crops

doi:10.1300/5895_06

MAP materials are used by a large number of industries such as pharmaceutical, food and spice, cosmetic, dental hygiene, paints, adhesive, beverage, herbal tea, agriculture, textile, paper and printing, motor, petroleum, plastic, tobacco, animal feed, dairy, and the like.

MEDICINAL AND PHARMACEUTICAL PURPOSES

MAP generate a number of useful compounds used in modern, allelopathic, and traditional medicines such as essential oils, alkaloids, terpenes, anthracenes, arbutrins, bitter principles, coumarins, flavonoids, glycosides, pungent principles, tannins, mucilages, saponins, hot substances, sweeteners, and the like. These compounds are isolated from plant materials by sophisticated extraction, isolation, fractionation, and purification techniques. The plant chemicals that are obtained are then used as phytochemicals or intermediates to produce modern drugs or traditional/herbal remedies. Some of the usual forms of processed plant products are powders, capsules, tablets, pills, teas, infusions, ointments, tinctures, extracts, syrups, injections, creams, lotions, standardized extracts, phytomedicines, and the like.

A large portion of the world population still relies mainly on traditional medicine. For example, in China, traditional medicine is widely used and accounts for 30 to 50 percent of total consumption (Zhang, 2001). Similarly, in India, especially in rural areas, 70 percent of the populations rely on traditional medicine (Zhang, 2001). In Germany, 50 percent of the plant-derived drug products are sold on medical prescription, the cost being refunded by health insurance (Gruenwald, 1997). About 80 percent of the populations in Benin and Sri Lanka use traditional medicine, and 60 to 70 percent of the rural population relies on traditional and natural medicine for their primary health care (Zhang, 1999). In the United States, the usage rate of the complementary/alternative therapies in 1997 ranged from 32 to 54 percent (Zhang, 2001). The use of plant-derived drugs is increasing in developing countries as well as developed countries. The demand for MAP is increasing due to the growing interest in natural products in place of synthetic chemical compounds, the increasing awareness regarding biodiversity conservation, sustainable, and protective use of plant resources, and the need for new pharmaceuticals for the prevention and cure of deadly diseases such as cancer and AIDS.

An estimate made by WHO shows that 80 percent of the population of developing countries rely on health care provided by plant-based medicaments. At least 25 percent of the drugs in modern pharmacopeias are still derived from plants; 121 such active compounds being in current use (Farnsworth and Soejarto, 1985; Posey and Overal, 1990; Rates, 2001). Of the 252 drugs considered as basic and essential by the WHO, 11.1 percent are produced from plants and a significant number are synthetic drugs obtained from natural precursors (Medeiros and Eraldo, 1999; Rates, 2001). Epipodophyllotoxin and podophyllotoxin from *Podophyllum peltatum,* camptothecin from *Camptotheca acuminata,* quinine and quinidine from *Cinchona* spp., codeine and morphine from *Papaver somniferum,* digitoxin from *Digitalis purpurea,* vincristrine and vinblastine from *Catharanthus roseus,* atropine from *Atropa belladonna,* and tubocurarine from *Chondrodendron tomentosum* are a few examples of the essential drugs isolated from plants with the help of traditional medicine (King, 1992; Cragg et al., 1993; Piesch and Wheeler, 1993; Baker et al., 1995; Balick and Cox, 1996).

Some of the pharmaceutical/therapeutic influences of the MAP can be indicated as on the blood system *(Panax quinquefolium),* menstruation *(Hydratis, canadensis),* and neural system *(Apium graveolens, Citrus bergamia)* regulators, and as antirheumatic *(Chamomilla recucita, Hydratis, canadensis, Taraxacum officinale),* antibiotic *(Allium, sativum, Aleo vera),* stimulant *(Borago officinalis),* antioxidant (Lavandula *officinalis, Melissa officinalis, Piper nigrum*), anti-inflammatory *(Maranta arundinacea),* digestive *(Pimpinella anisum),* analgesic *(Ocimum basilicum),* antiseptic *(Laurus nobilis),* anesthetic *(Atropa belladonna),* narcotic *(Papaver somniferum),* astringent *(Sytrax benzoin),* carminative *(Digitalis purpurea, Elettaria cardamomum, Tanacetum parthenium),* liver tonic *(Cichorium intybus),* emollient *(Malva sylvestris),* antiviral, *(Echinacea purpurea, Cinnamomum, zylanicum),* antidepressant *(Gingko biloba, Hypericum perforatum),* antihypertensive *(Crataegus laevigata),* antispasmodic *(Piper methsticum, Panax quinquefolium),* antifungal *(Origanum majorana, Rosmarinus officinalis),* sedative *(Passiflora incarnata),* anticancer (*Taxus* spp., *Silybum marianum*), expectorant *(Thymus vulgaris, Cinnamomum camphora),* and so on.

FOOD AND FOOD INGREDIENTS

Many of the medicinal and aromatic plants are edible and used as food, spice, seasonings, food flavorings, food colorants, and food preservatives in different countries and traditions. Recently, growing interests in green consumerism, ethnic foods, natural colorants, new foods (functional foods, nutraceuticals, and health foods), beverages and tastes (flavored coffee, instant herbal teas, nutritional and energy drinks) have increased the consumption of these plants (Anonymous, 1999a, 2001; Gruenwald and Herzberg, 2002).

Medicinal and aromatic plants are added to foods and beverages to add flavor, color, taste and health benefits, to enhance digestion, and to increase nutritional value and the shelf life of the foods. Whole herbs, young leaves, leaves, roots, rhizomes, shoots, flowers, fruits, barks, and seeds are the most frequently used plant parts. These plants are used as fresh, frozen, dried, and processed. Globally, fresh consumption of edible medicinal and aromatic plants is increasing due to the consumer demand for green consumption. However, the limited shelf life of the fresh products restricts the increased usage of these plants year around. Therefore, especially in developing countries, there is an urgent need for technological improvements in handling, storage, and transportation techniques for fresh products.

Most of the medicinal and aromatic plants are used in processed forms as dried, powdered, pills, capsules, tinctures, and so on, either as a single plant or as mixed preparations. Among the processed MAP products, functional foods have rapidly expanded their market size (Gruenwald and Herzberg, 2002). Functional foods are the foods that have health benefits with increased nutritional value by adding some bioactive ingredients, mostly phytochemicals (Gruenwald and Herzberg, 2002). On the other hand, fresh and processed MAP products are generally used in soups, salads, sauces, salad dressings, jams, jellies, cakes, candies, eggs, dips, vegetables, desserts, soft cheese spreads, butters, beef, lamb, fish, poultry, fruit cups, and so on to create new taste, to mask undesirable flavors, to give and increase flavor. Menthol, vanilla cocoa, several fruit juices, essential oils, herbs, spice, and oleoresins are the most widely used natural food flavors (Transfer et al., 1999; Anonymous, 1999a). The most popular flavors used in the soft drink sector are orange, lemon, lime, and apple (Anonymous, 1999a).

The turmeric extract (curcumin—bright yellow color), chlorophyll (alfalfa and several green plants—olive green color), caramel (sugar syrup—red/brown color), carotenoids (extracted from edible fruits, vegetables, flowers and fungi—colors from pale yellow through orange to red), β-carotene (extracted from carrots, *Dunaliella algae,* and palm oil—colors from yellow to orange), capsanthine and capsorubin (Paprica extract—colors from red to orange), xanthophylls lutein (extracted from dried ground flower petals of marigold and alfalfa—golden yellow color), and anthocyanin (extracted from black grapeskin and red cabbage—colors from orange/red to blue/red) are examples of plant-derived food colors/color sources that are permitted in the European Union (Transfer et al., 1999; Anonymous, 2000).

The common mallow, crab apple, parsley, peppermint, dill, horseradish, celery, garlic, onion, lemon, grapefruit, orange, oak fruits, thyme, fennel, dandelion, common nettle, rosehip, mushrooms, bay leaves, lemon basil, lemon grass, lemon thyme, rosemary, sorrel, borage, sage, lavender, marjoram, savory, tarragon, cinnamon sticks, clove, mace, chicory, cardamom, cassia, mustard, cumin, nutmeg, paprika, black pepper, edible amaranth, saffron, anise, caraway, poppy seeds, vanilla, sesame, ginger root, watercress, and chamomile are the MAP most widely used as food, spice, and condiments.

HERBAL TEA

Herbal teas are made from dried leaves, barks, flowers, flowering tops, berries, and seeds of medicinal and aromatic plants. In many traditions, it is believed that herbal teas have several beneficial effects like physical, mental, and emotional, and support general health with their bioactive constituents. In addition, herbal teas have their own therapeutic potent and flavor, depending on the constituents. Herbal teas can be made with a single plant (black or green tea) or mixtures of several plants, essential oils (essential oil of rose, lemon, etc.), aroma compounds (bergamot, cinnamon, cloves), and natural plant-derived colorants (hibiscus, rose hip, cinnamon) (Runham, 1996). Some of the medicinal and aromatic plants, for example, parsley leaves have diuretic effect in herbal teas, while peppermint is used to relieve colds.

In general, certain plant mixtures, single plant parts, essential oils, or whole herbs are used as herbal teas or herbal tea ingredients. The anise seed, alfalfa, angelica, bergamot, parsley, borage, caraway, fenugreek, hyssop, rose hip, blackberry, lemon balm, lemon geranium, feverfew, lemon grass, rosemary, chamomile, flaxseed, orange, sarsaparilla, chickweed, ginseng, peppermint, lemon verbena, lovage, marigold petals, savory, sage, chicory, gotu kola, psyllium, slippery elm, dandelion, hibiscus, raspberry, spearmint, echinacea, sweet woodruff, hops, marjoram, red raspberry, valerian, fennel, rosemary, horehound, thyme, cinnamon, black cohosh, buchu, buckthorn, burdock root, cascara sagrada, clover, cloves, cornsilk, damiana, dried pineapple and papaya, elderflowers, eyebright, ginger, elecampane root, gingko biloba, horsetail, lady's mantle, linden flowers, liquorice, marshmallow, meadowsweet, mother wort, nettle, vervain, oatstraw, passionflower, rhubarb root, saw palmetto, skullcap, willow herb, uva-ursi, wild yam root, St. John's wort, and yarrow are the most commonly used herbal teas or herbal tea materials (Anonymous, 1998, 2003a,b, 2004).

COSMETICS, PERFUMERY, AND AROMATHERAPY

As previously discussed medicinal and aromatic plants and their products have several therapeutic/pharmaceutical activities. As in the traditional medicines and aromatherapy, the cosmetic and perfumery industries also use the beneficial effects or properties (i.e., antimicrobial, anti fungal, anti-aging, anti bacterial, aromatic, antiseptic, sweetening, anti-irritation, relaxant, etc.) of these plants/plant products in their end-products. For example, flavor and fragrance industries use essential oils to create fragrance formulas that are then used in the production of several perfumes (Brud, 1995; Anonymous, 2000).

The cosmetic and perfumery industries use a broad range of *preprocessed* MAP products or plant-based raw materials such as essential oils (jasmine, peppermint, rose, etc.), tinctures, resins, gums, oleoresins, balsams, floral waters, absolutes, pomades, natural colors (indigo, carmine, marigold), concretes, extracts, standardized extracts, and the like. These products are then used as a raw/starting material for the production of a broad range of cosmetic and perfumery end-products like crèmes, lotions, tonics, perfumes, deodorants, colognes, shaving preparations, powders, bathroom products (soaps, shampoos, balsams), hair styling gels, antiperspirants, color cosmetics, oral hygiene products (toothpastes, mouthwashes, antiseptics), detergents, air fresheners, and sun, hair, facial, hand, skin and body care products (Zhaobang, 1995; De Silva, 1995; Runham, 1996; Anonymous, 1999a,b). Aloe vera, rosa damescena, calendula kava kava, lavender, ginger, eucalyptus, vetiver, basil, bergamot, jasmine, clove, lemon, orange, rosewood, eucalyptus, neroli, rosemary, thyme, ylang-ylang, sage, patchouli, peppermint, and chamomile are a few examples of the plants used in these industries.

On the other hand, consumer demands for natural cosmetic and perfume products are increasing all over the world. Recently, there is a new trend in consumer demands for the products that have therapeutic qualities namely *cosmeceuticals* to repair damaged tissues, smoothen, protect from the sun, and moisturize (Verlet, 1995; Anonymous, 2000). The production of these products requires several active ingredients having known biological effects, that is, extracts with anti-inflammatory activity *(Chamomilla recicuta),* sedative *(Hydrastis canadensis),* and wound healing properties *(Aloe vera, Echinacea purpurea, Chamomilla recicuta, Calendula officinalis).* Aromatherapy could be grouped under this trend due to the close relationship among the herbal medicine, perfumery, and aromatherapy (Runham, 1996; Stevensen, 1998; Anonymous, 2000).

Aromatherapy is the therapeutic use of plant-based essential oils to promote balance and harmony between physical, mental, and emotional conditions via massages, compresses, sprays, baths, showers, inhalation, perfume, and the like (Anonymous, 2000). As most of the essential oils are toxic, external use of the oils is suggested rather than internal use. *Rosa damescena* (dermatitis), *Eucalyptus citriodora* (anti-inflammatory, analgesic), *Jasminum officinalis* (antiseptic), *Juniperus communis* and *Eucalyptus radiata* (acne) and *Melissa officinalis* (viral skin infections) are the few examples of major essential oils used in dermatological treatments.

PEST AND DISEASE CONTROL

The use of synthetic chemical pesticides for pest and disease control leads to serious environmental and health problems. The use of known pesticides is no longer recommended, because of their long persistence and toxic residues in foods and several environmental concerns (Fishwick, 1988; Foegeding and Busta, 1991; Wright et al., 1993; Huang et al., 1997; Wang et al., 2001; Lee et al., 2001). Therefore, current researches are concentrated on the development of environmentally friendly natural pesticides derived from plants as an alternative to conventionally used synthetic pest and disease control agents.

Medicinal and aromatic plants provide powerful potential alternatives to currently used chemical pest and disease control agents because they contain numerous bioactive compounds, especially in the form of essential oils and extracts, having insecticidal, herbicidal, fungicidal, rodenticidal, antifeedant, antimicrobial, allelopathic, antioxidant, repellent, nematicidal, and bioregulatory properties (French, 1985; Caccioni et al., 1995a,b; Shaaya et al., 1997; Huang et al., 1997; Tripathi et al., 2000; Tunc and Erler, 2000; Keita et al., 2000; Lee et al., 2001; Papachristos and Stamopoulos, 2002; Huang et al., 2002; Kim et al., 2003). Nicotine, rotenone, and pyrethrins are a few examples of well-known commercially used plant-derived insecticides (Nivsarkar et al., 2001).

The essential oils and/or extracts of *Ocimum basilicum, Ocimum canum, Artemisa annua, rosemary, eucalyptus, garlic, nutmeg, Cinnamomum cassia, Cinnamomum sieboldii, Foeniculum vulgare, Pogostemon heyneanus, horseradish,* and *mustard* can successfully be used as insecticides against major stored-grain insects: *S. oryzae, S. zeamis, T. castaneum, C. Maculatus* (F.), *Callosobruchus chinensis* (L.), and *Stegobium paniceum* (L.) (Desphande et al., 1974; Desphande and Tipnis, 1977; Ho et al., 1996; Huang et al., 1997; Tripathi et al., 2000; Keita et al., 2000; Lee et al., 2001; Kim et al., 2003).

Nutmeg oil can be useful as a grain protectant with contact, fumigant, and antifeedant activities against *T. castaneum* (Herbst) and *S. zeamais* Motsch (Huang et al., 1997). The essential oils of oregano, thyme, basil, garlic, onion, and cinnamon are reported to be highly effective antimicrobials (Paster et al., 1995; Basilico and Basilico,

1999; Yin and Tsao, 1999). The essential oils of *Pimpinella anisum* (L.), and *Mentha piperita* (L.) have fumigant toxicity effects on the four major stored-product pests, namely, *R. dominica* (F.), *T. castaneum* (Herbst), *S. Oryzae* (L.), and *O. surinamensis* (L.) (Shaaya et al., 1991).

The essential oils of *Citrus funebris, Pinus sylvestris, Citrus tangerine, Citrus bergamia,* and *Eucalyptus citriodora* showed good repellent activity against *L. bostrychophila* (Wang et al., 2001). The growth and spore germination of *Aspergillus niger, Aspergillus ochraceus,* and *Aspergillus flavus* have been fully inhibited by the use of thyme (600 µg/mL) and oregano (400 µg/mL) essential oils (Paster et al., 1990). On the other hand, the major constituents of cinnamon and clove essential oils, cinnamic aldehydes and eugenol, have inhibited mold growth (Bullerman et al., 1977).

More examples of medicinal and aromatic plants used in pest and disease control can be found in Golop et al. (1999) where over 100 plant species used as bioactive protectants for grains are described in detail.

NONCONVENTIONAL USES

Among the known uses of MAP described above in detail, these plants are also used in the following manner:

- Dye in textile, painting, paper, and printing industries. Several weeds, wildflowers, lichens, mushrooms, trees, barks, roots, nuts, fruits, leaves, tannin extracts, rosins, and bryophytes are the sources of natural dyes; for example black/brown dyes from *Juglans nigra* tree; red dyes from yellow dock, St. John's wort, and madder; brown and grey color from blackberry and raspberry leaves; yellow dyes from black oak, weld, barberry, and sassafras; blue dyes from woad, indigo, and blueberry; and green dyes from horsetails (Simon et al., 1984; Blandrin et al., 1985; Thomas and Schumann, 1993; Zhaobang, 1995; Runham, 1996; Ogg, 1998; Anonymous, 1999b).
- Adhesives (gum arabic and gum ghatti), gelling agents (Agar-agar, starch, alginates, carrageenan, furcellaran, and pectin), tex-

turizers (locust bean gum and guar gum), emulsifying agent (gum karaya) (Anonymous, 1999b).

- Plant growth regulators (brassinolide from *Brassica napus*) and allelochemicals (terpenoids, coumarins, flavonoids, alkaloids, glycosides, phenolic acids, glucosinolotes, and tannins) obtained from different plants and plant essential oils (Blandrin et al., 1985; Putnam, 1988; Dudai et al., 1993; Runham, 1996; Bagchi et al., 1997). For example, *artemisinin and arteether* extracted from *Artemisia annua* are the phytotoxic agents inhibiting seed germination and seedling growth of various plants (Bagchi et al., 1997). In an another study, it was determined that, two sesquiterpene lactones (tagitinin A, tagitinin C) and a flavonoid hispidulin extracted from *Tithonia diversifolia* have inhibitory effects on the germination of radish, cucumber, and onion seeds (Baruah et al., 1994).
- Suppression sprouter (S-carvone from caravay used in potatoes, onions) (Palevitch, 1994; Runham, 1996).
- Papermaking (gum oleoresin, gum Arabic, guar gum, Manioc starch, rosins), rubber and plastic (gum oleoresin), petroleum, paint (rosins), and motor industries (Zhaobang, 1995; Anonymous, 1999b).

REFERENCES

Anonymous. 1998. Herbs and Herbal Teas. General Conference, Nutrition Council Andrews University Nutrition Department. Available from: http://www.andrews.edu/NUFS/herbs.html. Accessed 2003, September 20.

Anonymous. 1999a. Essential Oils and Oleoresins. A Survey of the Netherlands and other Major Markets in the European Union. Compiled for CBI by: Profound Adviser in Development, January 1999, 112 pp.

Anonymous. 1999b. Natural Gums and Resins. A survey of the Netherlands and other major markets in the European Union. Compiled for CBI by: ProFound, Adviser in Development, March 1999, 80 pp.

Anonymous. 2000. Natural Ingredients for Cosmetics. EU Market Survey, Compiled for CBI by: ProFound, Adviser in Development in collaboration with Mr. E-L. Roehl, June 2000, 49 pp.

Anonymous. 2001. Herb and Spice Market Assessment and Feasibility Study. Saskatchewan Agriculture and Food Innovation Consulting Group. Final Report. April, 2001, 228 pp.

Anonymous. 2003a. Herb Gardens for Different Cooking Styles, Available from: http://www.diabetic-lifestyle.com/articles/mar03_cooki_1.htm. Accessed 2003, October 02.

Anonymous. 2003b. Yogi Tea. Available from: http://www.yogitea.com/Herb Glossary/HerbGlossary.asp. Accessed 2003, June 11.

Anonymous. 2004. Herbal Teas. Tinderbox, Available from: http://users.highway1.com.au/~casandra/index.html. Accessed 2004, July 22.

Bagchi GD, Jain DC, Kumar S. 1997. Arteether: A potential plant growth inhibitor from *Artemisia annua. Phytochemistry;* 45(6): 1131-1133.

Baker JT, Borris RP, Carte B, Cordell GA, Soejarto DD, Cragg GM, Gupta MP, Iwu MM, Madulid DR, Tyler VE. 1995. Natural product drug discovery and development: New perspectives on international collaboration. *Journal of Natural Products;* 58(9): 1325-1357.

Balick MJ, Cox PA. 1996. *Plants, People and Culture: The Science of Ethnobotany.* New York, NY: Scientific American Library. 228 pp.

Baruah N C, Sarma JC, Barua NC, Sarma S, Sharma RP. 1994. Germination and growth inhibitory sesquiterpene lactones and a flavone from *Tithonia diversifolia. Phytochemistry;* 36(1): 29-36.

Basilico MZ, Basilico JC. 1999. Inhibitory effects of some spice essential oils on *Aspergillus ochraceus* NRRL 3174 growth and ochratoxin A production. *Letters in Applied Microbiology;* 29(4): 238-241.

Blandrin MF, Klocke JA, Wurtele ES, Bollinger WH. 1985. Natural plant chemicals: sources of industrial and medicinal materials. *Science;* 228(11): 54-60.

Brud WS. 1995. Formulation and Evaluation of Fragrance for Perfumery Cosmetics and Related Products. Chapter 5: 179-201, In: *A Manual on the Essential Oil Industry.* De Silva KT, Editor. Third UNIDO Workshop on Essential Oils and Aroma Chemicals Industries. Eskišehir, Turkey, November, 1995, Vienna, Austria, 232 pp.

Bullerman LB, Lieu FY, Seier SA. 1977. Inhibition of growth and aflatoxin production by cinnamon and clove oils, cinnamic aldehydes and eugenol. *Journal of Food Science;* 42: 637-646.

Caccioni DRL, Deans SG, Ruberto G. 1995a. Inhibitory effect of citrus fruit essential oil components on *Penicillium italicum* and *P. digitatum. Petria;* 5(2): 177-182.

Caccioni DRL, Ruberto G, Wang C, Rapisarda P. 1995b. Role of essential oil components on the resistance of citrus fruits to *Penicillium* spp. In: *Environmental Biotic Factors in Integrated Plant Disease Control.* Marka, M, Editor. 5-9 September 1994, Ponzan, Poland, p. 185-188.

Cragg GM, Boyd MR, Cardellina JH, Grever MR, Schepartz SA, Snader KM, Suffness M, 1993. The role of plants in the National Cancer Institute Drug Discovery and Development Programme. In: *Human Medicinal Agents from Plants.* Kinghorn, A.D. and Balandrin, M.F. Editors. ACS Symposium Series 534, Washington. pp. 80-95.

De Silva T. 1995. Development Programmes on Industrial Utilization of Aromatic Plants in developing Countries. United Nations Industrial Development Organization, Industrial Sectors and Environment Division. Third UNIDO Workshop on Essential Oils and Aroma Chemicals Industries. Eskišehir, Turkey, 6-9 November, 1995; p. 20.

Desphande RS, Adhikary PR, Tipnis HP. 1974. Stored grain pest control agents from *Nigella sativa* and *Pogostemon heyneanus. Bulletin of Grain Technology;* 12: 232-234.

Desphande RS, Tipnis HP. 1977. Insecticidal activity of *Ocimum basilicum* Linn. *Pesticides;* 11: 11-12.

Dudai N, Poljakoff-Mayber A, Lerner HL, Putievski E, Ravid U, Katzir I. 1993. Inhibition of germination and growth by volatiles of *Micromeria fruticosa.* In: *International Symposium on Medicinal and Aromatic Plants.* Palevitch D, Putievski E, Editors. Tiberias on the Sea of Galilee, Israel, March 22-25, 1993; 123-130.

Farnsworth NR, Soejarto DD. 1985. Potential consequence of plant extinction in the United States on the current and future availability of perception drugs. *Economy Botany;* 39(3): 231-240.

Fishwick FB. 1988. Pesticide residues in grain arising from postharvest treatment. *Aspects of Applied Biology;* 17: 37-46.

Foegeding PM, Busta FF. 1991. Chemical food preservatives. In: *Disinfections, Sterilization and Preservation.* Block SS, Editor. Lea and Febigert, Malvern, Pennsylvania, pp. 802-832.

French RC. 1985. The bio-regulatory action of flavour compounds on fungal spores and other propagules. *Annual Review of Phytopathology;* 23: 173-199.

Golop P, Moss C, Dales M, Fidgen A, Evans J, Gudrups I. 1999. The use of species and medicinals as bioactive protectants for grains. *FAO Agricultural services Bulletin,* Rome, Italy, 137: 239.

Gruenwald J. 1997. The market situation and marketing of herbal medicinal products (HMP) in Europe. In: *World Congress on Medicinal and Aromatic Plants for Human Welfare,* 2, 1997.

Gruenwald J, Herzberg F. 2002. The global nutraceuticals market. Business Briefing Innovative Food Ingredients; 7 pp.

Ho SH, Koh L, Ma Y, Huang Y, Sim KY. 1996. The oil of garlic, *Allium sativum* L. (Amaryllidaceae), as a potential grain protectant against *Tribolium castaneum* (Herbst) and *Sitophilus zeamais* Motsch. *Postharvest Biology and Technology;* 9(1): 41-48.

Huang Y, Ho SH, Lee HC, Yap YL. 2002. Insecticidal properties of eugenol, isoeugenol and methyleugenol and their effects on nutrition of *Sitophilus zeamais* Motsch. (Cleoptera: Curculionidae) and *Tribolium casteneum* (herbst) (Cleoptera: Tenebrionidae). *Journal of Stored Products Research;* 38(5): 403-412.

Huang Y, Tan JMWL, Kini RM, and Ho SH. 1997. Toxic and antifeedant action of nutmeg oil against *Tribolium castaneum* (Herbst) and *Sitophilus zeamais* Motsch. *Journal of Stored Products Research;* 33(4): 289-298.

Keita SM, Vincent C, Schmit JP, Ramaswamy S, Belanger A. 2000. Effect of various essential oils on *Callosobruchus maculatus* (F.) (Cleoptera: Bruchidae). *Journal of Stored Products Research;* 36(4): 355-364.

Kim SI, Roh JY, Kim DH, Lee HS, Ahn YJ. 2003. Insecticidal activities of aromatic plant extracts and essential oils against *Sitophilus oryzae* and *Callosobruchus chinensis. Journal of Stored Products Research;* 39(3): 293-303.

King SR. 1992. Pharmaceutical discovery, ethnobotany, tropical forest, and reciprocity: Integrating indigenous knowledge, conservation and sustainable development. In: *Sustainable Harvesting and Marketing of Rainforest Products.* Plotkin M, Famalarc L, Editors. Island Press, Washington, DC; 231-238.

Lee BH, Choi WS, Lee SE, Park BS. 2001. Fumigant toxicity of essential oils and their constituent compounds towards the rice weevil, *Sitophilus oryzae* (L.). *Crop Protection;* 20(4): 317-320.

Medeiros CN, Eraldo. 1999. Traditional use and sale of animals as medicines in Feira de Santana City, Bahia, Brazil. Indigenous Knowledge and Development Monitor. Available from: http://www.nuffic.nl/ciran/ikdm/7-2/medeiros.html. Accessed 2004, October 11.

Nivsarkar M, Cherian B, Padh H. 2001. Alpha-terthienil: A plant-derived new generation insecticide. *Current Science;* 81(6): 667-672.

Ogg KJ. 1998. Native dye plants of the United States. *Southern Illinois University Carbondale/Ethnobotanical Leaflets* Available from: http://www.siu.edu/~ebl/ Accessed 2003, May 13.

Palevitch D. 1994. Non-conventional uses of volatile oils and their constituents in agriculture. In: Proceedings of the 4th Symposium on the Economy of Medicinal and Aromatic Plants, 5-7 December 1994, Nyons; 26-40.

Papachristos DP, Stamopoulos DC. 2002. Repellent, toxic and reproduction inhibitory effects of essential oil vapours on *Acanthoscelides obtechus* (Say) (Cleoptera: Brucidae). *Journal of Stored Products Research;* 38(2): 117-128.

Paster N, Juven BJ, Shaaya E, Menasherov M, Nitzan, RH, Weisslowicz Ravid U. 1990. Inhibitory effects of oregano and thyme essential oils on mouldus and foodborne bacteria. *Letters in Applied Microbiology;* 11: 33-37.

Paster N, Menasherov M, Ravid U, Juven BJ. 1995. Antifungal activity of oregano and thyme essential oils applied as fumigants against fungi attacking stored grain. *Journal of Food Protection;* 58(1): 81-85.

Piesch RF, Wheeler NC. 1993. Intensive cultivation of Taxus species for the production of Taxol. *Acta Horticulture;* 344: 219-228.

Posey DA, Overal WL. 1990. *Ethnobiology: Implications and applications.* Proceedings of the First International Congress of Ethnobiology, Belem, Para, July 1988. Museu Paraense Emilio Goeldi, Belem.

Putnam AR. 1988. Allelochemicals from plants as herbicides. *Weed Tech;* 2: 510-518.

Rates SMK. 2001. Plants as a source of drugs-review. *Toxicon;* 39(5): 603-613.

Runham. 1996. An updated review of the potential uses of plants grown for extracts including essential oils and factors affecting their yield and composition (December, 1996), MAFF Ref: ST0105, ADAS Desk Study; 75 pp.

Shaaya E, Kostjukovski M, Eilberg J, Sukprakarn C. 1997. Plant oils as fumigants and contact insecticides for the control of stored-product insects. *Journal of Stored Products Research;* 33(1): 7-15.

Shaaya E, Ravid U, Paster N, Juven B, Zisman U, Pisarrev V. 1991. Fumigant toxicity of essential oils against four major stored-product insects. *Journal of Chemical Ecology;* 17: 499-504.

Simon JE, Chadwick AF, Craker, LE. 1984. *Herbs: An Indexed Bibliography. 1971-1980. The Scientific Literature on Selected Herbs, and Aromatic and Medicinal Plants of the Temperate Zone.*. Archon Books, Hamden, CT; 770 pp.

Stevensen CJ. 1998. Aromatherapy in dermatology. *Clin Dermatol;*16: 689-694.

Thomas MG, Schumann DR. 1993. *Income Opportunities in Special Forest Products. Self-Help Suggestion for Rural Entrepreneurs.* United States Department of Agriculture, Forest Service, Agricultural Information Bulletin 66, Washington D.C., May 1993; 206 pp.

Transfer HK, Lagemaat AJ, Oudenhoven L, Eshuis AF, van Leijden F. 1999. *Natural Food Colors and Flavours.* A compact survey of the Netherlands and other major markets in the European Union. Compiled for CBI, May 1999; 90 pp.

Tripathi AK, Prajapati V, Aggrawal KK, Khanuja SPS, Kumar S. 2000. Repellency and toxicity of oil from Artemisia annua to certain stored-product beetles. *Journal of Economic Entomology;* 93(1): 43-47.

Tunc I, Erler F. 2000. Fumigant activity of anethole, a major component of essential oil of anise *Pimpinella anisum* L. Integrated Protection of Stored Products, *IOBC Bulletin;* 23(10): 221-225.

Verlet N. 1995. Commercialization of essential oils and aroma chemicals (Chapter 6; 203-232). In: *A Manual on the Essential Oil Industry.* De Silva KT, Editor. Third UNIDO Workshop on Essential Oils and Aroma Chemicals Industries. Eskişehir, Turkey, November, 1995, Vienna, Austria; 232 pp.

Wang JJ, Tsai JH, Ding W, Zhao ZM, Li LS. 2001. Toxic effects of six plants oils alone and in combination with controlled atmosphere on *Liposcelis bostrychophila* (Psocoptera: Liposcelididae). *Journal of Economic Entomology;* 94(5): 1296-1301.

Wright CG, Leidy RB, Dupree J. 1993. Cypermethrin in the ambient air and on surface rooms treated for cockroaches. *Bulletin of Environmental Contamination and Toxicology;* 51: 356-360.

Yin MC, Tsao SM. 1999. Inhibitory effect of *Allium* plants upon three *Aspergillus* species. *International Journal of Food Microbiology;* 49(1-2): 49-56.

Zhang X. 1999. Traditional medicine worldwide–the role of the WHO. *Drug Information Journal;* 33(1): 321-326.

Zhang X. 2001. *Legal Status of Traditional Medicine and Complementary/Alternative Medicine: A Worldwide Review.* World Health Organization, 189 pp.

Zhaobang S. 1995. *Production and Standards for Chemical Non-Wood Forest Products in China.* Center for International Forestry Research, Occasional Paper No: 6, October, 1995; 18 pp.

Chapter 7

Decision Making

Sait M. Say

INTRODUCTION

In the last decade most of the agricultural production styles were not supported by public sources, so growers now have to compete in a global market. This trend is also valid for MAP production. The days of farming as a way of life are rapidly disappearing, and growers have to be aware of market demands. They also need to be well versed in preparing financial statements, forward planning, and making key management decisions. In this situation, growers must be business people first and growers second. As commodity prices continue to fall or level out, there is a great need for cost control to make as much profit as possible. So, a grower needs to know not only how to cultivate his crops but also how to manage his farm. At present, a new revolution is occurring in agriculture—the management revolution. Changes in agricultural technology, communications, transportation, management methods, data handling, human abilities, and the size and nature of farm-related industries have set the stage for this new era. A discussion of the role and functions of management in a changing and complex environment needs to begin with a definition of farm management. Perhaps a dictionary definition of management will provide a starting point. Following is one such definition:

I would like to thank the RIRDC (Rural Industries Research and Development Cooperation) for their useful publication, which is called Anonymous, 1999, in line with the details given in reference sections of chapter.

Medicinal and Aromatic Crops

doi:10.1300/5895_07

Management: The human factor within a production process which delimits problems, accumulates information relevant to their solution, analyzes that information, reaches decisions, acts on those decisions, and bears responsibility for the consequences of those actions (Herren and Donahue, 1990).

In parallel to this definition, farm management can be defined as the science and art of the organization and operation of farms so as to obtain the maximum amount of continuous net income. It considers the effectiveness of different sizes of operating units and of combinations of productive resources, enterprises, and practices for operating units; programs of adjustment for agricultural areas; and the impact of public policies and programs on economic activities and income on farms. There are some differences in the management of a farm and the management of a nonfarm business. The obvious differences are in size, type of business, and the products or services produced. Other differences include the relationship between labor and management and the setting of goals. According to these definitions and explanations, the general questions that have to be decided on are given as the following:

- How much to produce or how many inputs to use
- The most profitable combination of inputs to use
- The most profitable combination of products to produce

Decisions or answers are interrelated with the following issues:

- The physical relationships among inputs and outputs
- Prices of inputs and outputs
- Production economics criteria for determining the most profitable situation

In the light of this approach, the decision of new production selection or expanding the current activity is made according to financial and other relevant considerations. In the following, the decision-making process and detailed preference criterion are evaluated together to find the best solution for any grower. Finding the best solution is a complex process covering the decision-making rules.

DECISION-MAKING PROCEDURES

The firm represents the business or decision-making unit in an industry and involves the producer (or entrepreneur) in the manipulation of resources to obtain output. In farming, the farm is the firm and the farmer the manager or entrepreneur. The need to plan production arises from three basic factors:

- Individuals have various wants that they seek to satisfy.
- The means available to satisfy these wants are in scarce supply.
- The means available can be put to many different uses.

The planning problem is thus one of allocating scarce resources among various uses in a way that best satisfies the wants of the individual (Barnard and Nix, 1976). Successful growers have always identified the enterprises that offer them the best combination of profit, stability, and satisfaction for the resources available on their farm. Also, growers should continue to seek and find the next best alternative available to them. It may be an established enterprise with improved management practices or it may be a completely new information requirement, new marketing skills, new risks, and new management practices. The success of production planning and management is highly related to accurate sequential decision making.

Whatever the name given, the unique objective for an investor is "the analysis of the economics of business/engineering alternatives with the purpose of arriving at the best decisions" (Kavrakoglu, 1992). There is no doubt decision making is a complicated, vague, and difficult activity. The rapid economic growth in developed countries, international market expansion, diversification of products, developments in information technology, explosive penetration of new technology, and so on, all place a tremendous pressure on decision makers, both in the public as well as in the private realm. The competitiveness in the markets is forcing producers to be ever more careful in their actions. Economies open up to each other, whereby instabilities move easily across frontiers with greater speed. Currencies are susceptible to inflation or deflation, and exchange rates move quite violently at times. All this dynamism must be taken into account when making economic decisions.

On the other hand, by taking an analytical approach, adhering to established practices, and carefully observing certain procedures, decision making can be facilitated significantly. Of course, investment decisions in MAP production cannot be distinguished from this approach. The decision-making process is basically no different from the general problem-solving process. We can formalize it as follows (Kay, 1986; Kavrakoglu, 1992):

- Identify and define the problem.
- Collect relevant data, facts, and information.
- Establish objective(s) and criteria.
- Identify and analyze alternative solutions.
- Make the decision—select the best alternative.
- Implement the decision.
- Evaluate the results and bear responsibility for the outcome.

Providing that the general framework of the decision-making process presented above is considered, a specialization should be done for crop production decisions. The decision-making process for MAP production can be examined based on Figure 7.1, which is also valid for other fields of plant production. Figure 7.1 presents a two-staged procedure, namely decision basics and application and decision check. In the decision basics step, the criterions for crop selection among alternatives are given. In the second stage, application and decision check process regarding management issues is presented.

In this chapter, decision basics around the crop selection issue are discussed. While financial implications and financial evaluation methods are evaluated in detail, the other important investment decision factors are briefly explained.

FACTORS FOR NEW ENTERPRISE DECISION

Through the years, MAP growers have contended with weather, insects, crop diseases, financial challenges, global competition, changes in government regulations, and a variety of other obstacles that increase the risk of MAP production. Improving profits and prolonging production are always the primary concerns for growers. Policy makers and growers are often seeking a miracle crop enterprise that

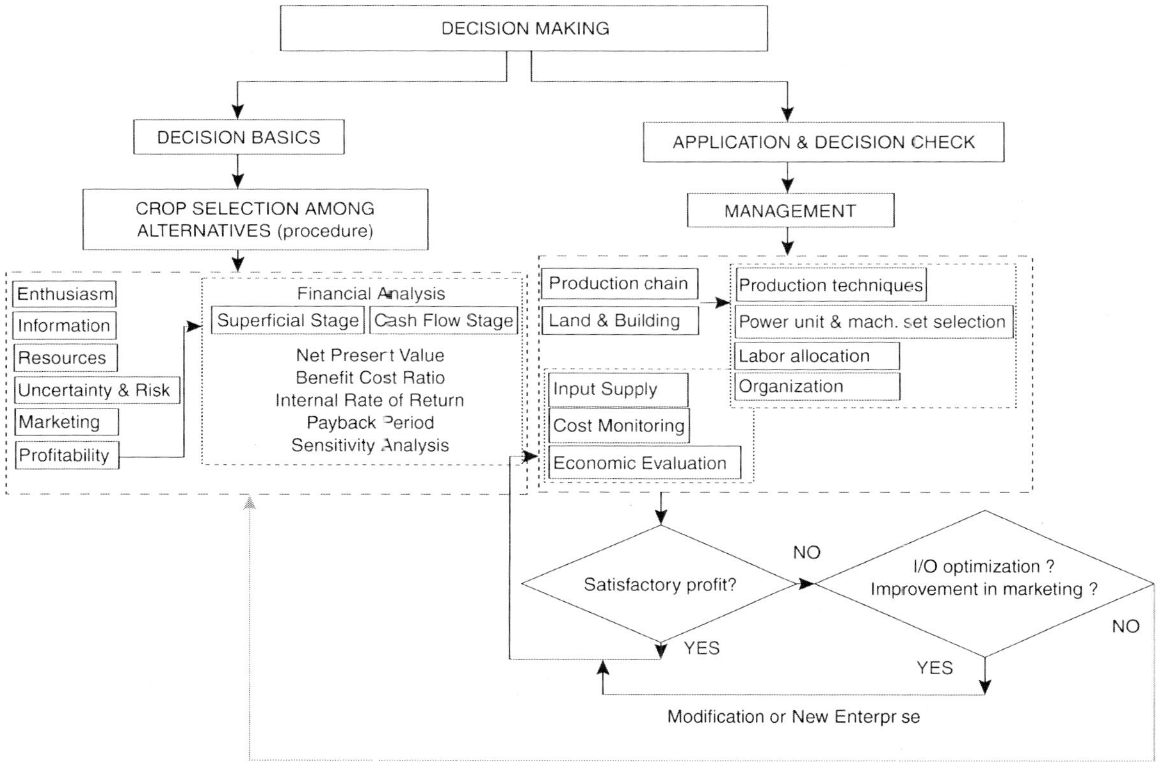

FIGURE 7.1. Decision-making procedure.

will bring immediate wealth. The search for what to produce has led many growers to experiment with new and different crop opportunities with varying degrees of success (Woods and Isaacs, 2000). Perhaps the major concern should be on the process of *right enterprise selection*. It is not easy to guide in this manner. These important questions are satisfactorily answered linked with the decision-making process. In the process, the specific nature of MAP production and the behaviors of the grower should be taken into account. On the other hand, from growers to government officials, decision makers around the world are confronting a range of challenging issues in agriculture, as rapid changes in technology, domestic policies, demand patterns, international trade rules, and environmental conditions affect the organization and performance of MAP production at all levels. When one decides on a new or expanded enterprise, a high possibility of different sets of financial, marketing, management, and policy preferences are encountered. There is a strict relationship between grower behavior, with its effecting factors, and new enterprise decision. These factors are the basic elements that should drive any decision to adopt a new enterprise on the farm. On the other hand, focusing on the selection of the right enterprise within the alternatives will allow the entrepreneur to thoroughly evaluate a wide range of options. Beyond deciding on which crop is more profitable among the feasible alternatives for agricultural production, when any grower selects the medicinal and aromatic plant production with an expected profit, expectations can be tested by considering the appropriate data and methods.

In the following sections, driving factors based on the Woods and Isaacs's approach, with some modifications for the new investment in agricultural production, are listed and then explained (Woods and Isaacs, 2000):

1. Enthusiasm/Motivation
2. Information
3. Resources
4. Uncertainty and Risk
5. Marketing
6. Profitability

Enthusiasm/Motivation

Investment in MAP production is partly considered as a subject that falls under behavioral science. It is also governed by personnel characteristics such as conformity to past customs, group behavior, and flexibility to change. Enthusiasm and personnel characteristics have considerable effects on production decisions. It is well known that enthusiasm and motivation are interrelated. Motivation has been defined as "all those inner-striving conditions described as wishes, desires, drives, etc. It is an inner state that activates or moves" (Henrietta, 2007). From a manager's perspective, a person who is motivated (Donnelly et al., 1995):

- works hard,
- sustains a pace of hard work, and
- has self-directed behavior toward important goals.

Thus, motivation and enthusiasm involve effort, persistence, and goals. They involve a person's desire to perform.

Enthusiasm perhaps could also stand for emotion or entrepreneurial ability. In the optimism of planning a new enterprise, it may be difficult to think of reasons for getting out of business. The grower's judgment may be clouded by the investment of time, money, and emotion that has gone into the enterprise (Woods and Isaacs, 2000). Enthusiasm should be justified by short-, intermediate-, and long-term goals for the enterprise, and these goals should be reviewed at least annually to assess progress or to allow for revision. In some cases, growers look for conformity to past customs for production practices of new crops. Most growers who investigate alternative production areas agree that a quick changing ability is very important. As the nature of MAP growing techniques evolves, the way market-oriented production is conducted and the ability to respond to opportunity is becoming increasingly important. But, today, changing the current production route is long, wearisome, and risky in most farms. It is seldom well planned, organized, or controlled. In many, if not most, cases it just happens. As a result, change is not viewed as a friend to new market-oriented MAP production. It is feared and avoided. In fact, conformity to past customs and flexibility to change in any agricultural activity should be considered together. If enthusiasm, which was supported by reasonable and prudent analysis, is greater than the sum of effects of the frustrating factors, decision on new crop production is likely to be positive. As a result, enthusiasm conflicting with frustrating factors is an initiative for detailed evaluation.

Information

Managing production and operations to achieve high levels of productivity and quality requires information. Information is really a fuel that drives organizations. A major purpose of a manager is to convert information into action through the process of decision making. The required information includes cost data, projected revenues, setup times, lead times, input/parts requirements, labor availability and capability, machine capabilities and capacities, vendor reliability, and competitor intelligence.

There is a strict interrelationship between information and other driven factors. Risk-bearing capacity and available resources should optimally be matched with adequate and sufficient information for good results in production. Quality information is critical for making good decisions and evaluations of any existing enterprise. Good information about unusual alternatives may be scarce and expensive. The more unusual the enterprise the less likely that conventional sources will be able to meet the need for technical or economic information. An inventory of useful information sources for a particular enterprise can become a valuable resource in its own right and can greatly aid in ongoing management decisions (Woods and Isaacs, 2000). A simple procedure needed to turn the information into knowledge and practice is presented in Figure 7.2.

Key information subjects for an agricultural production activity can be grouped as follows:

- Production information (books, extension publications, magazines, internet, face to face interviews, field trips etc.)
- Financial information (balance sheets, income statements, cash flow budgets, etc.)
- Marketing information (potential outlets, input suppliers, price trends, etc.)
- Human resources (consultants, specialists, attorneys, web sites, internet connections etc.)
- Educational opportunities (trade shows, extension meetings, short courses, etc.)

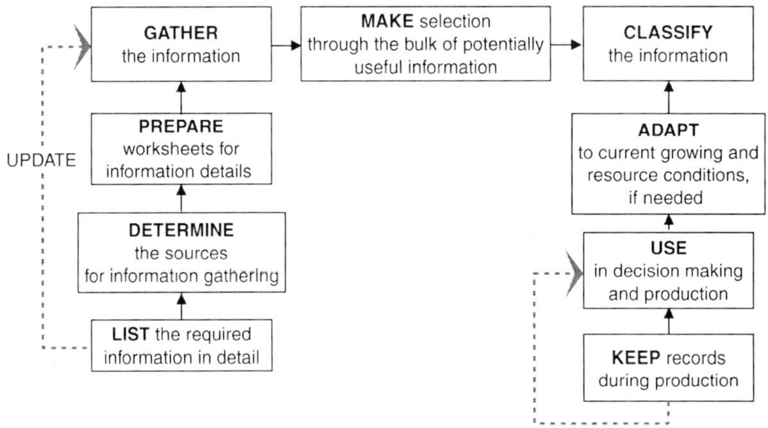

FIGURE 7.2. Simple information management procedure.

A wide range of specialized information and expertise can be provided by using these information channels around the world. Owing to the huge number of information sources, classification is recommended. After classification of the information, useful information should be derived from the bulk. In some cases, data adoption or transformation may be needed by considering the current growing and resource conditions. Then, classified and adopted information is ready to use in calculations for decision making and production. Keeping farm records is another activity essential to make an information pool from real conditions for future decisions and traceability. Also, an information pool can be useful for other producers. On the other hand, cost and reliability of the information gathered is also to be considered. Besides, the use of all the information mentioned above can bring in success, especially financial and marketing implications have a main effect on the do or don't decision. In the following paragraphs, investment decision is discussed from the marketing and financial point of view in detail.

Resources

A manager must consider the resources available for attaining the goals that have been set. Limits are placed on goal attainment because most managers are faced with a limited amount of resources.

In a farm or ranch business, goal attainment is confined within some limits set by the amount of land, labor, and capital available (Kay, 1986).

Undoubtedly, new or expanded resources are needed to achieve the new or expanded production goal. New selections for different kinds and capacities of machinery, buildings, and/or land should be made regarding the both short- and long-term plans and expectations. A procedure from requirement determination to the decision on purchasing or hiring of any resource item is presented in Figure 7.3.

The resources mentioned may change over time, but they are never available in infinite amounts. Identifying current resource limits and acquiring additional resources, including management skills, are continual problems confronting a farm manager.

Uncertainty and Risk

Because of the time lag in agricultural production and our inability to predict the future accurately, there are varying amounts of risk and uncertainty in all farm and ranch management decisions (Kay, 1986).

Risk is defined as something with an uncertain outcome. You make a decision today when you don't know what the future will

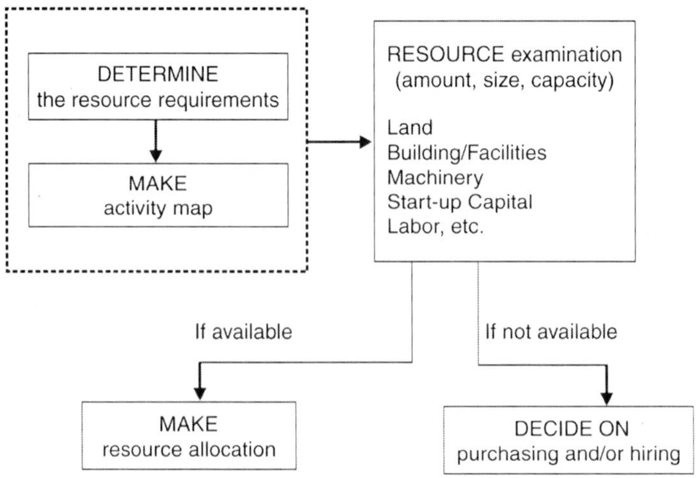

FIGURE 7.3. Resource allocation procedure.

bring. There is a casual relationship between uncertainty and risk; uncertainty refers to factors not under control and not known with certainty. Whereas risk is a hazard because of uncertainty (Ku, 2003).

Farm managers find their best decisions often turn out to be less than perfect because of changes that have taken place between the time the decision was made and the time the outcome of that decision is finalized or known. Many agricultural decisions have outcomes months or years after the initial decisions were made. Crop farmers must make decisions on crops to be planted, seeding rates, fertilizer levels, and other input levels early in the cropping season. The crop yield obtained as a result of these decisions will not be known with certainty for several months or even several years in the case of perennial crops.

As is generally known, risk and agricultural production go hand in hand. Land earns rent, labor earns wages, management earns a salary, capital earns interest, and *assuming risk in MAP production earns profit.* Without risk in agriculture there would be no chance of profit. All risk will ultimately affect income. This may include production, financial, or marketing risk (Figure 7.4). It is well known that new emerging enterprises or some modifications on the current one in any farm usually carry considerably different kinds of risks. MAP production, which will possibly be faced by many different risks, can be evaluated as a new emerging enterprise. Several of them are common to all agricultural production such as production and technical risk, price or marketing risk, and financial risk.

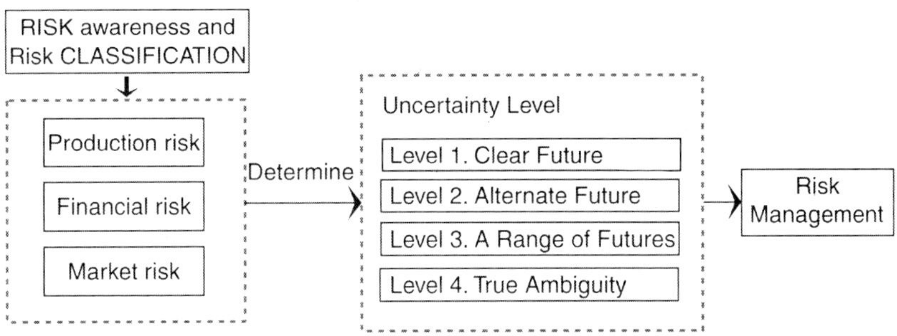

FIGURE 7.4. Risk management and decision procedure.

Crop yields are not known with certainty before harvest or the final sale. Weather, diseases, insects, and weeds are examples of factors that cannot be accurately predicted and can cause yield variability. These yield variations are an example of production risk. Another source of production risk is new technology. There is always some risk involved when changing from an old and proven production technology to some new technology.

Risk in MAP production is often considered as having specific characteristics. A major source of risk in agriculture is the variability of output prices. Commodity prices vary from year to year and many have substantial seasonal variation within a year. Growers are not well informed on species that are frequently demanded in the international markets. On the other hand, cultivation techniques of many MAP species are not studied on a scientific base. There is a lack of extension services in this field. Collected information on specific topics is not shared within the growers or countries because of the secret of commercial interests. Agricultural inputs, especially mechanization techniques could not be easily obtained from the market, which is normally no problem for other market crops.

Financial risks are related to the amount of capital borrowed, loan terms (length, interest rate, periodic payment), liquidity, solvency, and profitability (Rabinovicz, 2002). Market risks such as seasonal price variability, seasonal high and low prices, perishability of product, and food safety concerns should be derived from historic data and trends. Uncertainty levels for each risk item are listed as follows with their brief explanations (Courtney et al., 1997).

Level 1 (Clear Enough Future)

At level 1, managers can develop a single forecast of the future that is precise enough for strategy development. Although it will be inexact to the degree that all business environments are inherently uncertain, the forecast will be sufficiently narrow to point to a single strategic direction.

Level 2 (Alternate Futures)

At level 2, the future can be described as one of a few alternate outcomes or discrete scenarios. Analysis can not identify which outcome will occur, although it may help establish probabilities. Most important, some, if not all, elements of the strategy would change if the outcome were predictable.

Level 3 (A Range of Futures)

At level 3, a range of potential futures can be identified. This range is defined by a limited number of key variables, but the actual outcome may lie anywhere along a continuum bounded by that range. Most of the agricultural risks are in this group.

Level 4 (True Ambiguity)

At level 4, multiple dimensions of uncertainty interact to create an environment that is virtually impossible to predict. Unlike in level 3 situations, the range of potential outcomes can not be identified. It may not even be possible to identify, much less predict, all the relevant variables that will define the future. Level 4 situations are quite rare.

Riskier enterprises require different risk management strategies. Low debt loads, diversification, irrigation, insurance, engaging in contract marketing, credit reserves, leasing equipment, joint purchasing of inputs and marketing of products, and hedging on futures markets are only a few strategies that should be considered in evaluating and adopting new enterprise. It should be remembered that the more unusual or obscure the enterprise, the riskier it will be.

Marketing

Marketing can be defined as a social and managerial process by which individuals and groups obtain what they need and want through creating, offering, and exchanging products of value with others (Anonymous, 2001). Marketing should be evaluated with special interest, especially, for the riskier production branches that will ultimately be affected by all risks like agriculture. The marketing process must be designed not only to evaluate the current status for decision but to follow the ever changing environment of the market. Marketing management is accomplished by carrying out marketing research, planning, implementation, and control. A generalized guide covering marketing issues for growers is shown in Figure 7.5.

Marketing research is the first step of the whole process. This stage is likely the key to the success or failure of any new enterprise. So, growers need relevant, accurate, reliable, valid, and current data to complete the marketing research stage. As a whole, required information can be gathered from a private consultancy service or governmental institution. In the planning stage of the marketing issue, SWOT (Strengths, Weaknesses, Opportunities, Threats) analysis should be used. SWOT analysis is a very effective way of identifying a farm's strengths and weakness, and of examining the opportunities and threats faced (Anonymous, 2002). Carrying out an analysis using the SWOT framework helps the grower to focus his activities into areas where are strong and where the greatest opportunities lie. SWOT analysis can be used to identify and analyze the strengths and weaknesses of your farm as well as the opportunities and threats revealed by the information you have gathered on the external environment. It

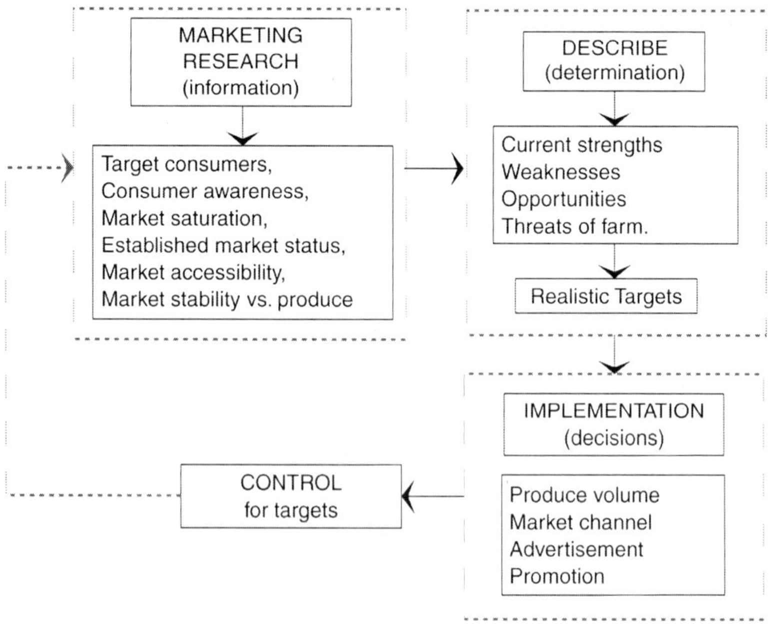

FIGURE 7.5. General guide for marketing issues.

helps to make a decision or develop a plan that takes into consideration many different internal and external factors, and maximizes the potential of the strengths and opportunities while minimizing the impact of the weaknesses and threats. SWOT analysis considerations and marketing research results are combined to determine the realistic targets. Outcomes from these targets lead to profit estimations. Marketing implementation is also critical to success. In fact, targets lead to determination of the implementation procedure. For example, produce volume and market channel are the main selection items around targets. Alternative market channels (wholesale, retail, coops, contracts, and direct sales) should be compared by means of product and market characteristics. To compare the alternatives, information about the number of grower transactions per year, marketing fees (for storage, packaging, handling and delivery, advertising, insurance, licensing/legal fees, membership fees, and others), prices (historical averages at least for three to five years, lower bound, upper bound), market reliability (growth potential), and value-adding opportunities (additional customer services, special packaging, delivery, processing, preserving, special promotion and others) are required from different sources. Decision on the marketing channel is especially important. In this context, it can be remembered that each channel has different benefits and cost patterns. In addition, market saturation level and profit potential of a given channel should be evaluated in detail. Prices received by the grower often vary widely across marketing channels. It is a fact that the opportunities for new markets can be created by partnership with other producers. The implementation stage is highly related to market research.

The last step is control of marketing targets. Because of the likely variability in production year by year, marketing channels and capacity can be changed. In such a case, cost/benefit calculations should be revised at the end of each year. Calculations for this stage are interrelated with profitability estimations, as will be discussed in the next paragraphs. According to results, all marketing processes should be questioned.

Continuous product improvement, building the marketing effort around valued product features, and having a well-defined market plan may be the difference between the ultimate success and failure of the enterprise (Woods and Isaacs, 2000).

Profitability

Agricultural enterprise requires a great deal of analysis and planning, whether the new production is to be an independent operation or an addition to an existing farm. Making a decision on an agricultural production is possibly more complicated than other investment areas because there are so many risk features, as mentioned earlier, to be understood and evaluated for clear views regarding specific agricultural conditions. When a grower decides on a new production area, questions arise concerning the evaluation process of the decisions on a financial basis and the selection of tools and techniques to be used. Financial analysis, which investigates the profitability, is one of the most important processes in decision making. Profit maximization is a widely accepted goal, particularly as it contributes to other potential goals such as the growth and survival of the business.

The financial model of a situation is far less complicated than the situation itself. If there were no such concept as money, making a decision would be difficult: Each alternative leads to a set of estimated future consequences, and one would have to compare one multidimensional description with another. With money and the techniques of engineering economy, each set of estimated future consequences has a worth expressed as a single real number, and the comparison is one dimensional (Young, 1993). The decision paradigm mentioned can be executed within a financial analysis framework. Building a framework leading to step-by-step evaluation for financial analysis should be the main interest.

FINANCIAL ANALYSIS

Financial implications of any new and/or expanding investment describe the profitability based on some realistic estimations and calculations. Beyond the crop selection principles mentioned previously, profitability is the main valuation criterion. And most importantly, there is a mutual interaction between profitability and the other principles. Profitability, solvency, and liquidity are the three important goals of any production planning. As a goal, profit is not always understood well. Sometimes it is confused with cash flow. Sometimes it is confused with the highest income or the lowest cost

(Kriegl, 1992). In rough terms, profitability can be identified by deducting various kinds of expenses from total estimated revenue.

Clearly any new or expanding production must add to the overall profits of the farm. New enterprise should be budgeted closely with estimates of costs and returns. In other words, when making a financial analysis before investment, all possible cost and returns related to profitability should be considered in detail. After the design of the profitability-oriented decision-making procedure (Figure 7.6) all data should be gathered from reliable sources and by consistent estimations.

Deciding on the duration and size has its main effects on the results of the financial analyses. The duration is completely selected according to the grower's views, based on his own expertise and foreseeing ability. High foreseeing ability is related to knowing about the history of the agricultural policies, marketing status, prices, and production variability, and estimating the global and country-wide changes.

Generation of Alternatives

Every decision problem has a scope, which is a definition of the limits of the problem, limits as to which kinds of alternatives can be considered. Generation of alternatives is only valid in the event of the existence of two or more hopeful investment areas. The procedure

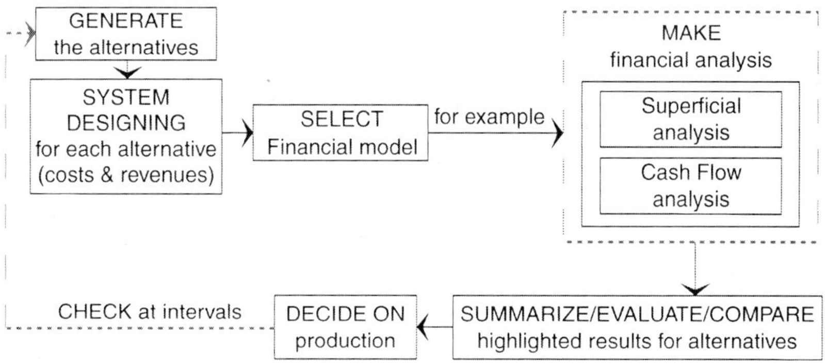

FIGURE 7.6. Profitability-oriented decision-making process.

presented in Figure 7.6 can be either applied to all possible alternatives or only to a selected one. Each alternative should be analyzed in a logical and organized manner to ensure accuracy and to prevent something from being overlooked. After the determination of outcomes from each alternative, results should be compared to select the best one.

System Design for Costs/Revenues

After the generation of alternatives, a checklist should be prepared for all cost and revenue components. This checklist is also used as a guide in the Superficial Stage and Cash Flow Analysis calculations and production chain design that is included in the management phase of enterprise.

Superficial Stage

In a sense, the new or expanded enterprise is evaluated for holding an opinion in its first look feasibility in superficial stage. All necessary data should be compiled and designed in such in a way that there should not be any lack of clarity for using them in proposed financial analysis. On the other hand, the accuracy of data design for the possible investment under question is the other important factor for accurate results. Data design prevents the overlooking of any cost or revenue item, which leads to reduction in profitability. The cost of a wrong decision means that for one set of assumptions, there is an optimum position and net revenue function that permits computing the reduction in net revenue due to nonoptimum positions, or use of different (i.e., wrong) assumptions. For the proposed financial analysis, which will be discussed in following paragraphs in detail, is widely accepted for almost all kinds of investment interrogations. When designing the data, *investment inputs, seasonal inputs, yield,* and *demand* are itemized in detail. Table 7.1 presents all the fundamental components of a MAP production as an example enterprise (Anonymous, 1999).

The input portion of the model is divided into the four parameter sections, namely investment inputs, seasonal inputs, yield, and the

TABLE 7.1. General identification for all fundamental components of MAP.

Component	Description
Investment Inputs	
Field survey	Topography, soil type, agronomic characteristics, farm planning (field roads, buildings, warehouse, workshop, etc.)
Stock	Seed stock/seedlings etc.
Machinery	All required mobile and stationary farm machineries
Establishment	Fencing, land forming, banks, channels and drains, trellises, access structures
Processing	Drying, distillation, fermentation, etc.
Distribution	Packing facilities, bulk silos, bulk transporters, reefer trucks
Markets	Established or prospective markets
Permits, etc.	Permits, licenses, industry association memberships
Working capital	Normally, value of one season's inputs
Seasonal Inputs	
Reproducing material	Seed, rootstock, cuttings
Irrigation/water	Irrigation if necessary
Soil preparation	Seed and/or seedlings bed preparation, stubble burning
Fertilizers	Phosphates, urea, etc.
Chemicals	Herbicides, pesticides
Crop protection	Trellises, different kinds of covers, etc.
Harvesting	Crop harvesting
Repair and Maintenance	Fencing, banks, structures, machinery
Disposal	Disposal of unused products (e.g., grape skins, grain husks)
Permits, etc.	Annual permit, license, industry levy renewals
Yield	
Primary yield	Unit yield of primary crop
By-product yield	Unit yield of each by-product
Demand	
Demand value	Unit farm-gate price, or market price minus transport and handling
Quantified demand	Size of accessible market
Price elasticity	Impact of new supply on market prices
Projected demand	Market outlook

Source: Adapted from Anonymous, 1999. Used with permission from RIRDC.

revenues/demand. All inputs to complete the analysis framework are given in Table 7.2. After the general identification of all fundamental components of MAP and calculation requirements, the principles of the financial analysis should be introduced.

TABLE 7.2. All required inputs for financial analysis.

Inputs	Requirement
Investment Inputs	Appropriate unit of production (e.g., hectare), and reasonable size of enterprise First time cost of all investment items Expected useful life of each in years Real interest rate assumption (see Equation 7.1) Annualized cost calculation of each item (see Equation 7.2) Total cost calculation Calculation of the total investment and annualized cost
Seasonal Inputs (price/unit)	Estimation of all the input costs Total cost calculation
Yield (kg/ha)	Estimation of the anticipated yield for primary crop Estimation of the anticipated yield for all by-products
Revenues (price/kg)	Estimation of the anticipated farm gate price for the primary crop Estimation of the anticipated farm gate price for all by-products Total revenues

Beyond the data design and input parameters the main sections of Superficial Stage analysis are

- model results and
- sensitivity and threshold analysis (proposed in RIRDC report; see Anonymous 1999).

Model results and sensitivity and threshold analysis are explained in detail in the following paragraphs. One of the most important basic data for the analysis is the unit of production (e.g., a hectare of crop) selected by the grower. Superficial analysis steps are diagrammatically summarized in Figure 7.7.

The annualization of the investment costs is computed on an annuity basis. The interest rate input by the user is taken as a discount rate (real interest rate) for this calculation. The real interest rate is the key factor for the calculation of annualization for input parameters.

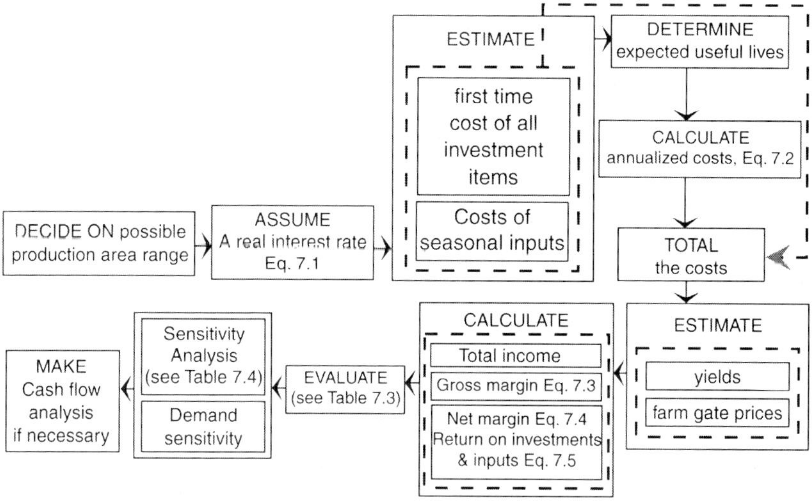

FIGURE 7.7. Steps for the superficial analysis.

Real interest rate is called "inflation-free" interest rate, according to the following basic relationship (Young, 1993):

$$i_r = \frac{1 + i_m}{1 + i_j} \qquad (7.1)$$

where

i_r = real interest rate (discount rate), *decimal;*

i_m = market interest rate or the combined interest rate, *decimal;* and

i_j = annual inflation rate, *decimal.*

Calculations of the annualized cost of each investment input are made according to the following equation (Gonen, 1990):

$$A = P\left[\frac{i_r (1 + i_r)^N}{(1 + i_r) - 1}\right] \qquad (7.2)$$

where

A = annualized cost of given input, €, \$ etc.;
P = present cost of given input, €, \$ etc.; and
N = useful life of given input, years.

Model Results

In the results section, the gross margin (Equation 7.3) and the net margin (Equation 7.4) for the enterprise for an almost typical growing season are calculated. It is also computed as an approximation of the enterprise's net return on seasonal costs (RSC) (Equation 7.5) alone as well as on investment and seasonal costs (RISC) (Equation 7.6). This procedure is called superficial analysis.

$$\text{Unit Gross Margin} = \text{Revenue} - \text{Recurrent Input Costs} \tag{7.3}$$

$$\text{Net Margin} = \text{Gross Margin} - \text{Annualized Investment Cost} \tag{7.4}$$

$$\text{Return on Seasonal Costs} = \frac{\text{Gross Margin}}{\text{Recurrent Costs}} \tag{7.5}$$

$$\begin{array}{c}\text{Return on Investment}\\ \text{and Seasonal Cost}\end{array} = \frac{\text{Net Margin}}{(\text{Revenue} - \text{Net Margin})} \tag{7.6}$$

These computations are more important for identifying the sensitivity of the enterprise's viability to the various data categories than for establishing its financial viability. Superficial analysis results are not sufficient to decide on the profitability of the enterprise. Evaluation of the superficial analysis results can be made according to some performance indicators presented in Table 7.3.

TABLE 7.3. Performance indicators for evaluation.

Performance indicators	**Main condition**	**Sub-Condition**	**Comment**
Net Margin and RISC	Net Margin > 0	RISC < 0*	Unfavorable outcome
		0 <RISC < 0.25	Favorable outcome
		RISC > 0.25	Very favorable outcome

Source: Anonymous, 1999. Used with permission from RIRDC.

Sensitivity and Threshold Analysis

Gross and net margin's sensitivity to the changes in revenue, seasonal and investment inputs is one of the other leading indicators for decision on new investment. For this purpose threshold analysis is performed over gross and net margin according to the criterions presented in Table 7.4. Threshold analysis gives some results which emphasize the importance level of the various data categories. By considering this guide, resource improving possibilities which will effect all investment costs and marketing issues may be evaluated.

Market Inquiry

As stated before, marketing is a key factor for an investment decision and is evaluated in financial analysis. For determination of the strengths and the weaknesses on marketing ability, some critical questions should be answered by considering the current market affinity. For example,

- established market existence,
- market accessibility possibilities,
- estimations on the effects of new supplies to market stability, and
- possible new market opportunities in the near future.

TABLE 7.4. Threshold analysis procedure.

Evaluation	Calculation (%)		Decision criterions	
"*zero*" GM	Revenue reduction (RR)	$= \frac{GM}{RR} \times 100$	RR,RIC;IIC>20%	not sensitive
			10%<RR,RIC; IIC<20%	sensitive
	Cost increment on recurrent inputs (RIC)	$= \frac{GM}{RIC} \times 100$	RR,RIC;IIC<10%	very sensitive
"*zero*" NM	Revenue reduction	$= \frac{NM}{RR} \times 100$		
	Cost increment on recurrent inputs	$= \frac{NM}{RIC} \times 100$		
	Cost increment on investment inputs (IIC)	$= \frac{GM}{IIC} \times 100$		

When answering questions like those mentioned and more, beyond the mathematical models and other calculations, past experiences which are in, even if, different areas should be taken into account.

Cash Flow Stage

Identification of prospects that are worthy of commercial advancement is comprehensively made in a detailed financial analysis by using engineering economy calculations. Whether the examined investment comes to true or not is supported by detailed financial analysis. So, a detailed financial analysis must take a wider and more complete set of data and view of the prospect than is appropriate in the superficial analysis. All components of a detailed analysis are summarized in Figure 7.8.

The main outcome of the cash flow analysis is the final decision on the declining or approving the possible investment under question.

The cash flow analysis is structured around a horizontal cash flow layout with annual time steps and a selected evaluation period. A twenty year evaluation period is commonly considered to be the maximum period over which a new enterprise can be evaluated (Anony-

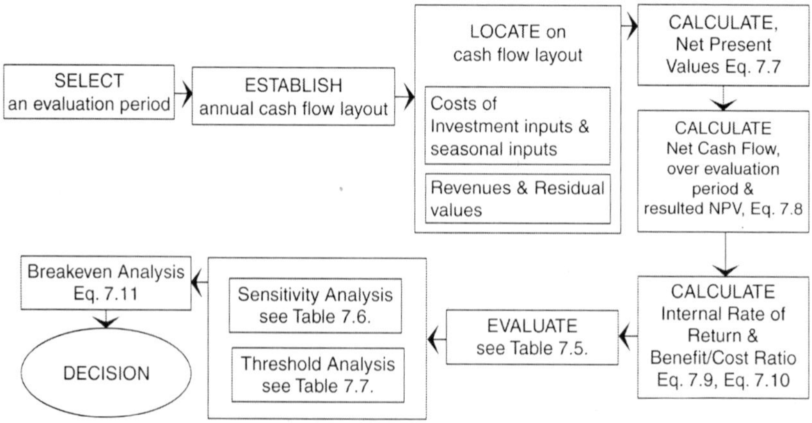

FIGURE 7.8. Steps of cash flow analysis.

mous, 1999). All decision alternatives involve elements of money, time, and risk, which are combined by means of certain procedures so that they can be compared with each other or within some acceptable ranges (Kavrakoglu, 1992). In financial analysis, it is considered a transfer of money to be characterized by its direction, its amount, and its timing after some arbitrary time zero. The direction is expressed by making the amount a signed number, with a positive sign (+) donating an inflow or receipt and a negative sign (−) denoting an outflow or disbursement (Young, 1993). As the name suggests, cash flow takes place whenever cash or its equivalent (e.g., check, transfer through bank accounts, or some other means) "flows" from one party to another (Gonen, 1990).

Fundamental decision criterion in this analysis is net present value or worth. The present value method converts all revenues or costs, that is, *cash flows,* representing an investment alternative into *equivalent worth* at some point or points in time using an interest rate equal to the minimum attractive rate of return (discount rate).

In the present worth method, each individual cash flow is converted to its value equivalent at the present, that is, the beginning of period 1 or the end of period 0, and then individual present worths are summed. Present value calculation is often called *discounting.* Therefore, the present value of a given investment alternative can be expressed as

$$PV(i) = \sum_{t=0}^{N} \frac{A_t}{(1+i)^t} \quad \text{or} \quad PV(i) = \sum_{t=0}^{N} A_t (1+i)^{-t} \quad \text{or}$$
$$PV(i) = \sum_{t=0}^{N} A_t (P/F, i\%, t) \tag{7.7}$$

where

PV(i) = present value of alternative, €, $, etc.;
A_t = cash flow at the end of period t, €, $, etc.;
N = study period, years;
P = present value of money; and
F = future value of money.

If the alternatives are mutually exclusive, they may be ranked according to their net present values. The best alternative is the one with the highest net present value. If the investment alternatives are independent, then all alternatives with net present values greater than zero satisfy the discount rate criterion and, therefore, are economically justifiable.

Net cash flow calculations during the evaluation period are made by summing up the cash flows of all items of investment inputs, recurrent inputs, revenues, and residuals for each year. On the other hand, the resulting net present value (*rNPV*) which is one of the most important evaluation criteria can be calculated from Equation 7.8.

$$rNPV = NPV_{ii} + NPV_{r} + NPV_{R} + NPV_{rs} \qquad (7.8)$$

The other evaluation criterion, namely internal rate of return (*IRR*), is the most sophisticated of financial metrics and is often used to analyze large, multiyear investments. IRR equals the percentage rate by which you have to discount the net benefits for your time period until the point at which they equal the initial costs. The rate of return calculated by IRR is the discount rate you would need to apply to your benefits to obtain a net present value of zero (Işik, 1999) (Equation 7.9).

$$PV = 0 = \sum_{t=0}^{N} \frac{A_t}{(1+i^*)^t} \qquad (7.9)$$

where i* (IRR) is the particular value of *i* that satisfies the equality condition.

The ratio of benefits to costs (B/C) is an efficiency measure that is conceptually simple and is versatile (Equation 7.10). It measures cost efficiency.

$$B/C = \frac{\sum_{t=0}^{N} R_t (1+i)^{-t}}{\left(\sum_{t=0}^{N} II_t (1+i)^{-t} + \sum_{t=0}^{N} RI_t (1+i)^{-t}\right) - \sum_{t=0}^{N} S_t (1+i)^{-t}} \qquad (7.10)$$

where

R_t = revenue at t, €, \$, etc.;
II_t = investment input at t, €, \$, etc.;
RI_t = recurrent input at t, €, \$, etc.; and
S = salvage value of an investment item at t, €, \$, etc.

After the computation of financial performance indicators, comments that lead to a decision should be made according to Table 7.5.

Sensitivity analysis includes the NPV of the enterprise if the discount rate is changed to be 2 percentage points higher and lower than that entered by the analyzer, and IRR with revenues, investment inputs, and seasonal recurrent inputs each adjusted by 10 percent. Table 7.6 gives comments on the evaluation of sensitivity analysis.

C* is calculated as

$$C^* = \left| \frac{\left(\sum_{t=0}^{N} NCF_t (1+(i \pm 0.02))^{-t} - \sum_{t=0}^{N} NCF_t (1+i)^{t} \right)}{\sum_{t=0}^{N} NCF_t (1+i)^{-t}} \right| \tag{7.11}$$

where NCF is net cash flows each year during the evaluation period.

TABLE 7.5. Comments about performance indicators.

Performance indicators	Condition	Comment
NPV	>0.10 of investment inputs	Very favorable outcome
	0 < NPV < 0.10 of investment inputs	Favorable outcome
	≤0	Unfavorable outcome
IRR	>1.5 × required rate of return	Very favorable outcome
	1 < IRR ≤ 1.5 × required rate of return	Favorable outcome
	< required rate of return	Unfavorable outcome
B/C Ratio	>1.5 1	Very favorable outcome
	1 < B/C Ratio ≤ 1.5	Favorable outcome
	≤1	Unfavorable outcome

Source: Anonymous, 1999. Used with permission from RIRDC.

TABLE 7.6. Comments for sensitivity analysis.*

Performance indicators	Condition	Comment
NPV	C* (Eq. 7.11)>1	Very sensitive
	C > 0.5	Sensitive
	C < 0.5	Not sensitive
	>0.2	Not sensitive
IRR	> 0.1	Sensitive
	≤0.1	Very sensitive

*Yield/prices decreased by 10 percent, investment expenditure increased by 10 percent, and recurrent inputs increased by 10 percent.

Threshold analysis results show how much each of the main parameter groups needs to change, either increase or decrease, to derive a zero value NPV. Comments for threshold analysis results are given in Table 7.7.

$$C_1 = \left| \frac{\sum_{t=0}^{N} NCF_t (1+i)^{-t}}{\sum_{t=0}^{N} R_t (1+i)^{-t}} \right| \tag{7.12}$$

$$C_2 = \left| \frac{\sum_{t=0}^{N} NCF_t (1+i)^{-t}}{\sum_{t=0}^{N} II_t (1+i)^{-t}} \right| \tag{7.13}$$

$$C_3 = \left| \frac{\sum_{t=0}^{N} NCF_t (1+i)^{-t}}{\sum_{t=0}^{N} RI_t (1+i)^{-t}} \right| \tag{7.14}$$

Threshold analysis is useful in assessing whether an unviable enterprise is reasonably likely to become viable with further investigation and improvement, or whether the change in a key assumption necessary to convert a viable enterprise to becoming unviable is a reasonable expectation.

Finally, the breakeven point along the time scale located in the analysis period is computed on a cumulative discounted basis. Table 7.8. gives the solution method for breakeven point analysis. The first year number that gives the positive CDCF is the breakeven point for the investment.

The last step is summarizing the highlighted results of the analysis. Key assumptions such as enterprise scale, initial investment, typical recurrent input costs, key yield factors, prices, discount rate, and analysis period should also be emphasized within the highlighted results. On the other hand, financial analysis results for both the superficial and detailed stage are denoted in a clear way. Covering all kinds of major risks will be cautionary and useful.

TABLE 7.7. Threshold analysis evaluation.

Performance indicators	**Condition**	**Comment**
NPV equals ZERO with		
yield/pricing decreasing by C_1% (Eq. 7.12)	C1,C2 and C3≤0.10	Very sensitive
investment expenditures increasing by C_2% (Eq. 7.13)	0.10<C1,C2 and C3≤0.20	Sensitive
recurrent inputs increasing by C_3% (Eq. 7.14)	C1,C2 and C3>0.20	Not sensitive

TABLE 7.8. Method of breakeven point analysis.

Years, *n*	**$CDCF_n$**	**Comparison parameter**
1	$NCF_n(1+i)^{-n}$	
2	$CDCF_{n-1}+NCF_n(1+i)^{-n}$	
.	.	$CDCF_n>0$
.	.	
.	.	
N	$CDCF_{N-1}+NCF_N(1+i)^{-N}$	

For the desired results and profit, all details should be evaluated by considering the dynamic environment of agricultural production. Because both crop and farm have some specifics, every new enterprise should be identified in detail. Successful growers have always identified the enterprises that offer them the best combination of profit, stability, and satisfaction for the resources available on their farm. Successful growers will continue to seek and find the next best alternative available to them. It may be an established enterprise with improved management practices or it may be a completely new enterprise with new information requirements, new marketing skills, new risks, and new management practices. The key is to know how to match resources and targets with logical and applicable approaches. By doing that, related former results based on scientific researches are also considered. It is to be remembered that the grower's happiness highly depends on converting the soil fertility to profit by scientific rules.

SOFTWARE FOR DECISION MAKING

The use of computer software in the decision-making processes is tremendously increasing in many industries. A large number of agricultural applications have been reported in the literature. All these computer-based decisions systems carry out their tasks performing two major functions; namely, (1) information management and (2) information manipulation (Lal et al., 1989). The control of inflow and outflow of information is the subject of information management. On the other hand, information manipulation is the process of working on the gathered information according to the captured knowledge to produce results. Farmers are increasingly purchasing and using on-farm computers to provide decision-support information. According to a report revealed by Nuthall (2004), some farmers have clearly found computer operation useful and perhaps the process of methodically collecting, entering, and interpreting data has a synergistic effect on mental decision making.

Of course, it can be difficult to fully understand and to apply the whole procedure of decision making for selecting the most profitable crop production activity. So, a software, named FANE v 1.0 (*Financial Analysis of New Enterprise*-Version 1.0), was developed to facilitate the financial evaluation of crop production enterprises. It is designed to develop an understanding of the start-up planning of an enterprise by using beforehand results, which are obtained by applying engineering economy concepts on detailed cost and benefit estimations. For this purpose, all costs and benefits should be estimated over the planned useful life of the enterprise in a proposed framework. Costs are categorized as investment and seasonal inputs. Benefits cover primary crop and by-product income. Briefly, the comprehensive economical feasibility of a given crop's growing activity can be examined by running FANE v 1.0. More detailed information on the software is presented in the Appendix.

REFERENCES

Anonymous 1999. *The New Rural Industries: Financial Indicators.* A Report for the Rural Industries Research & Development Corporation by Hassal & Associates. Edited by Keith Hyde. RIRDC Publication No:99/38. 54 pp.

Anonymous 2001. Importance of Marketing Research. Council of American Survey Res. Organizations. Available from: http://www.casro.org/media/Importance of Research.pdf (Accessed 2004, November 22).

Anonymous 2002. Swot Analysis. New Zealand Trade & Enterprise. Available from: http://www.marketnewzealand.com/common/files/swotanalysis-cl.pdf (Accessed 2004, November 26).

Barnard, C.S., Nix, J.S. 1976. *Farm Planning and Control.* Cambridge University Press, Great Britain, 549 pp.

Courtney H., Kirkland J., Viguerie P. 1997. Strategy under uncertainty. *Harvard Business Review.* November-December; 75(6): 66-79.

Donnelly, J.H., Gibson, J.L., Ivancevich, J.M. 1995. *Fundamentals of Management.* Richard D. Irwin, Inc., USA, 719 pp.

Gonen T. 1990. *Engineering Economy for Engineering Managers.* A Wiley-Interscience Publication. 455 pp.

Henrietta F. 2007. Motivation. Budapest University of Technology and Economics: Department of Management and Corporate Economics. 7 pp. Available from: http://www.mvt.bme.hu/imvttest/segedanyag/58/motivation_02.doc. Accessed 2006, November 22.

Herren R.V., Donahue R.L. 1990. *The Agriculture Dictionary.* Delmar Publishers Inc. 553 pp.

Işik A. 1999. *Engineering Economy* (in Turkish). Bizim Buro Basımevi, Ankara. 317 pp.

Kavrakoglu I. 1992. *Decision Economics.* Publication of Graduates Association from Bogazici University, Istanbul, Turkey. 263 pages.

Kay R.D. 1986. *Farm Management. Planning, Control, and Implementation.* McGraw-Hill, Inc. 401 pp.

Kriegl T. 1992. What Is Your Level of Profitability? Agri-Business, Cooperative Extension Service Publications, University of Wisconsin-Extension. Available from: http://cdp.wisc.edu/pdf/Profitlevels.pdf. Accessed 2004, November 22.

Ku A. 2003. *Modeling Uncertainty in Electricity Capacity Planning*. Thesis, London Business School. 241 pp.

Lal, H., Peart, R.M., Jones, J.W., Shoup, W.D. 1989. An Intelligent Information Manager for Knowledge-Based Systems. ASAE Paper No: 89-7083.

Nuthall, P.L. 2004. Case studies of the interactions between farm profitability and the use of a farm computer. *Computers and Electronics in Agriculture;* 42: 19-30.

Rabinovicz E. 2002. Should Risk Management Tools Play A Role in the CAP?. Swedish Institute for Food and Agricultural Economics. 16 P. Available from: http: // www.arl-net.de/west _ european_working_group/agriculture/rabinowicz .pdf (Accessed 2004, November 22).

Woods T., Isaacs S. 2000. *A PRIMER for Selecting New Enterprises for Your Farm.* Cooperative Extension Service, University of Kentucky, College of Agriculture. 24 pp.

Young D. 1993. *Modern Engineering Economy.* John Wiley and Sons, Inc. 563 pp.

Appendix

User-Friendly Software for Decision Making: FANE v 1.0

A software, named FANE v 1.0 (*Financial Analysis of New Enterprise-Version 1.0*), was developed for the financial evaluation of crop production enterprises. The software program was written in the C++ language. It is designed to develop an understanding of the start-up planning of an enterprise by using beforehand results, which are obtained by applying engineering economy concepts on detailed cost and benefit estimations. For this purpose, all costs and benefits should be estimated over the planned useful life of the enterprise in a proposed framework. Costs are categorized as investment and seasonal inputs. Benefits cover primary crop and by-product income. Briefly, the comprehensive economical feasibility of a given crop's growing activity can be examined by running FANE v 1.0.

More detailed information about the use of FANE v 1.0 is given in simple diagrams and output screens of the software, as shown in this chapter. Figures A.1 and A.2 depict the program flow and the flowchart of the software respectively.

START SCREEN

When FANE v 1.0 is started, two main buttons that activate the **New Project** and **Help** screens appear (Figure A.3). The **Help** screen contains two submenus, named **General** and **New Project**. In the **General** window, a brief description of the financial analysis components is given. The **New Project** window in the **Help** screen contains detailed definitions and explanations about **New Project** items, which will appear under **Data Entry** screen when the **New Project** button is clicked from the **START** screen.

Medicinal and Aromatic Crops

doi:10.1300/5895_08

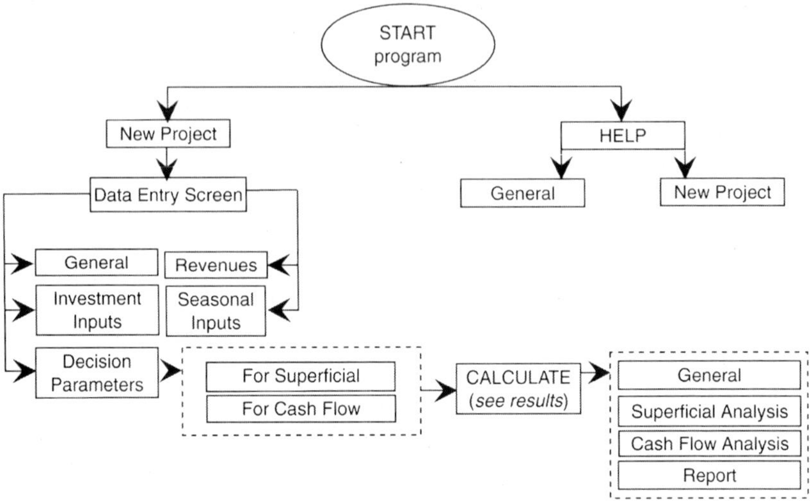

FIGURE A.1. Diagrammatical exhibition of process for FANE v 1.0.

NEW PROJECT SCREEN (DATA ENTRY SCREEN)

When the **New Project** button in the **START** menu is clicked, the **Data Entry** screen is displayed (Figure A.4). The **Data Entry** screen leads to five subscreens, each of them with data input cells. The names of these subscreens are **General, Investment Inputs, Seasonal Inputs, Revenues,** and **Decision Parameters.**

GENERAL SCREEN

In the **General** screen, some basic descriptive information and data are entered. Also, inputs and results of each analysis can be stored in the database as a separate file. A data set or a file is added as a file or deleted from the database by clicking the appropriate button located on just below the information entry section. It is also possible to access the previously stored files from the **General** screen. The discount rate is entered as a single projected value or is calculated by the software as a function of market interest and inflation rate (Figure A.4).

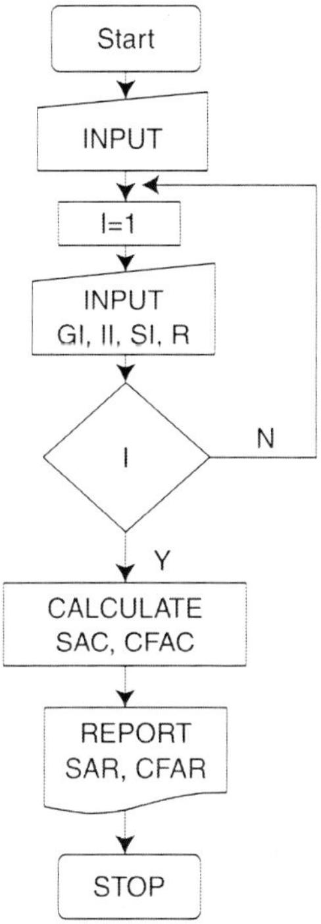

Nomenclature for flowchart:

AN	Investment analysis number
GI	General information (descriptive)
II	Investment inputs
SI	Seasonal inputs
R	Returns
DP	Decision paramoters
SAI	Superficial anaysis components
CFAI	Cash flow analysis components
SAR	Superficial analysis results
CFAR	Cash flow analysis results

FIGURE A.2. Flowchart of FANE v 1.0.

INVESTMENT INPUTS SCREEN

Investment inputs for any agricultural crop enterprise are entered in the software by filling in the cells under the six columns with the headings Cost Item, Detail, Number, Unit Charge, Useful Life, and Salvage Value (Figure A.5).

SEASONAL INPUTS SCREEN

The required seasonal inputs and their costs are entered in the software by adding the related data items in the cells under the five columns with the headings, Cost Item, Detail, Cost Cycle, Number or Amount per ha, and Unit Charge (Figure A.6).

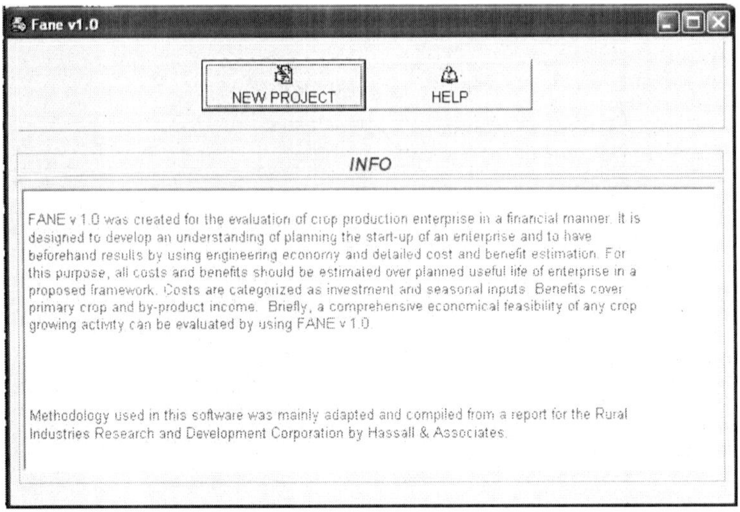

FIGURE A.3. The Start screen of FANE v 1.0.

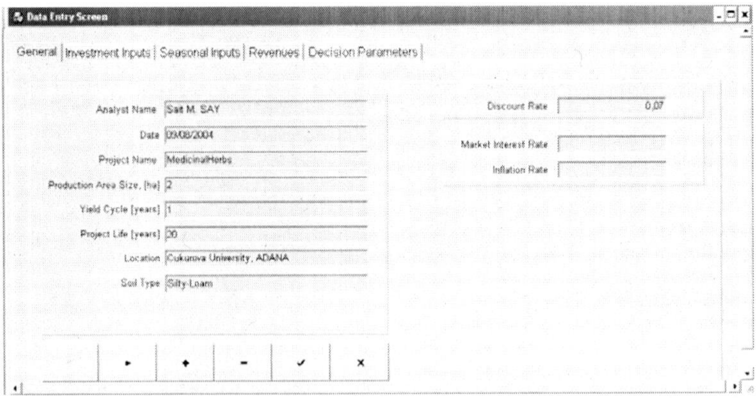

FIGURE A.4. Data Entry screen.

REVENUE/S SCREEN

All possible revenues from an agricultural enterprise are entered in the software under the three columns with the headings, Income Item, Estimated Yield, and Estimated Sale Price (Figure A.7).

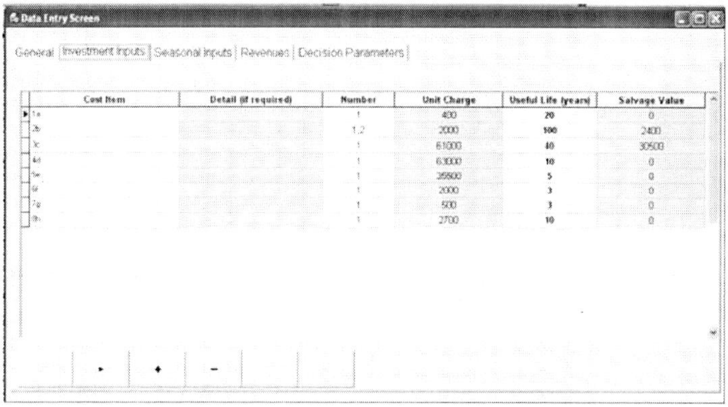

FIGURE A.5. Investment inputs screen.

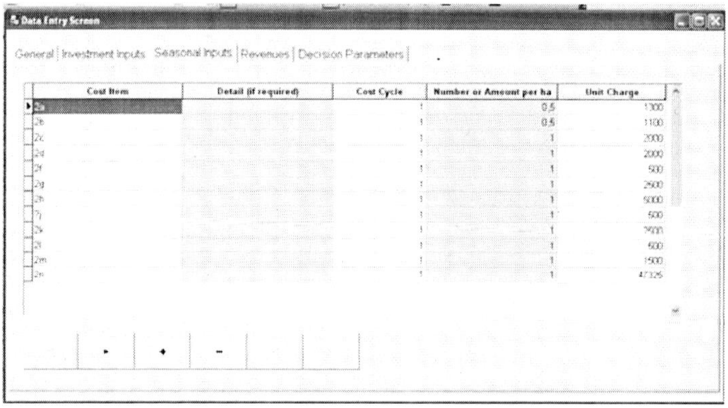

FIGURE A.6. Seasonal inputs screen.

DECISION PARAMETERS SCREEN

Some reference values are added to the software during the preparation stage of the financial analysis in the Decision Parameters screen for both superficial and cash flow analysis. These reference values, called decision parameters, are then compared with the calculated values of a given project to decide on the feasibility of the proposed enterprise (the details of the calculated values and comparison procedure are given in Help context of the software). The Decision Parameters screen is shown in Figure A.8.

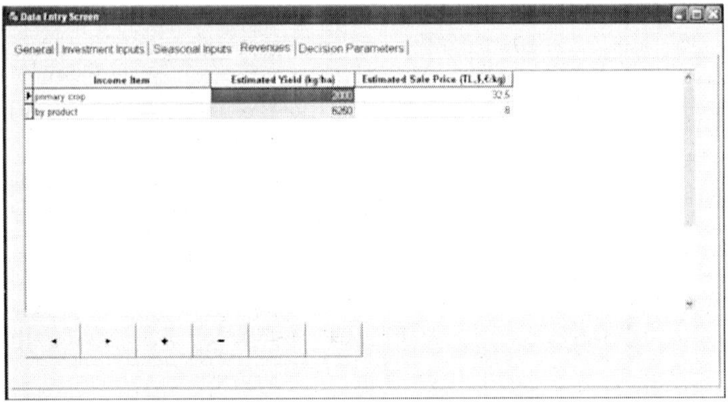

FIGURE A.7. Revenue/s screen.

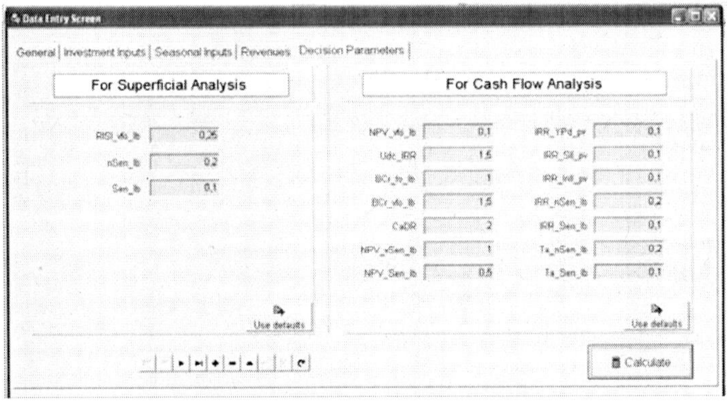

FIGURE A.8. Decision Parameters screen.

Based on the results of the comparisons between decision parameters and calculated values, two main comments are presented:

- An indication for favorability rating of the outcome of the given project such as *very favorable outcome, favorable outcome, unfavorable outcome*.
- An indication of the sensitiveness of the calculated values of a given project to the decision parameters, as *very sensitive, sensitive,* or *not sensitive*.

All lower–upper bounds of the decision parameters may vary according to the crop, economic conditions, and the profit expectation. Determination of upper–lower bounds requires expertise. In the case of lack of expertise, default values can be used for crop production. After the entry of the decision parameters, the **CALCULATE** button is clicked.

RESULTS SCREEN

The Results screen is divided into four sections, namely: General, Superficial Analysis, Cash Flow Analysis, and Report. The Superficial Analysis screen has two main parts, namely Cost–Revenue Results and Result Evaluation (Figure A.9). Investment Costs, Seasonal Costs, and Revenues are evaluated separately under the Cost–Revenue Results part. In the Result Evaluation part, Gross Margin and Net Margin of the given project are evaluated according to the related decision parameters entered in the Decision Parameters screen.

The Result screen of the Cash Flow Analysis was designed to contain two sections, namely CFA Results and Result Evaluation. The CFA Results screen contains five main subsections (Investment Costs, Seasonal Costs, Revenue, Salvage Values, and Net Cash Flow) (Figure A.10). All cost and

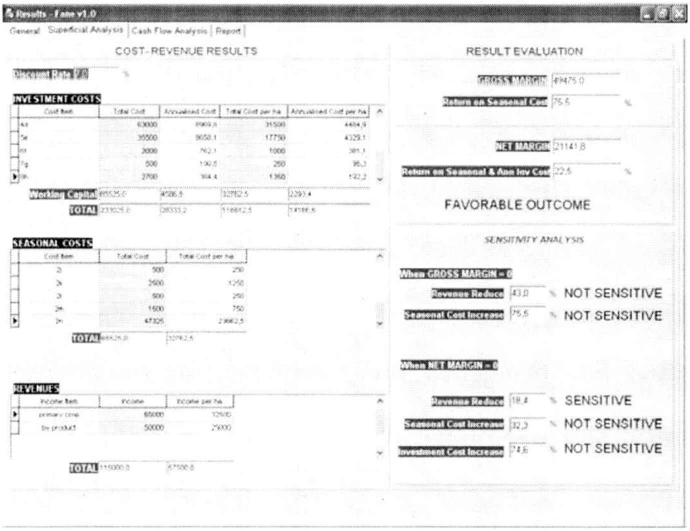

FIGURE A.9. Results screen for Superficial Analysis.

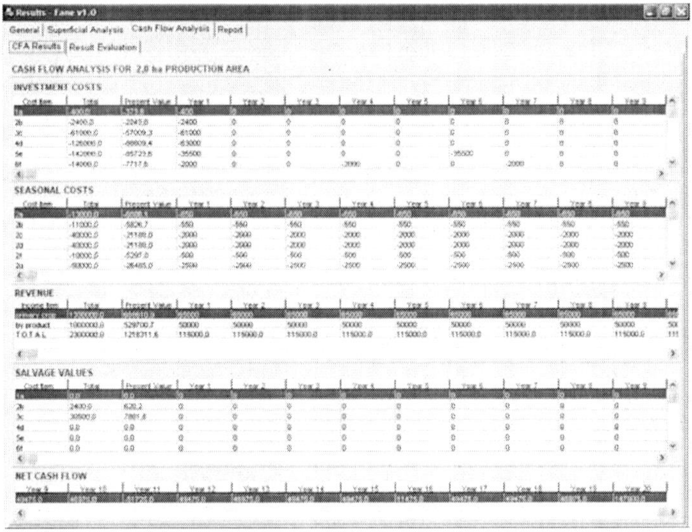

FIGURE A.10. CFA Results screen.

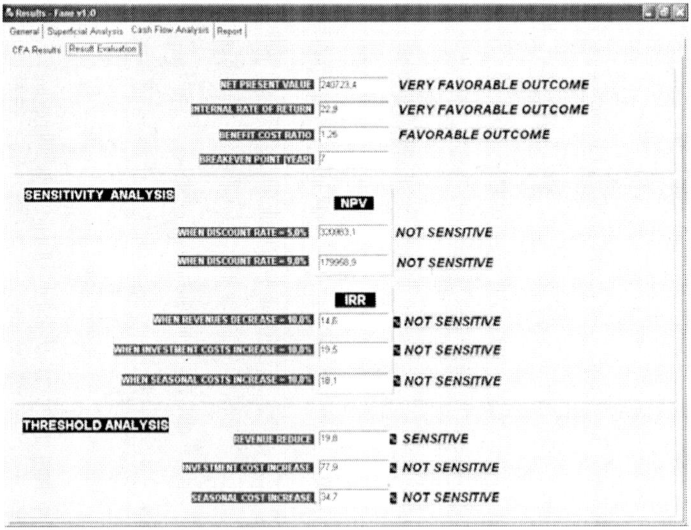

FIGURE A.11. Result Evaluation screen.

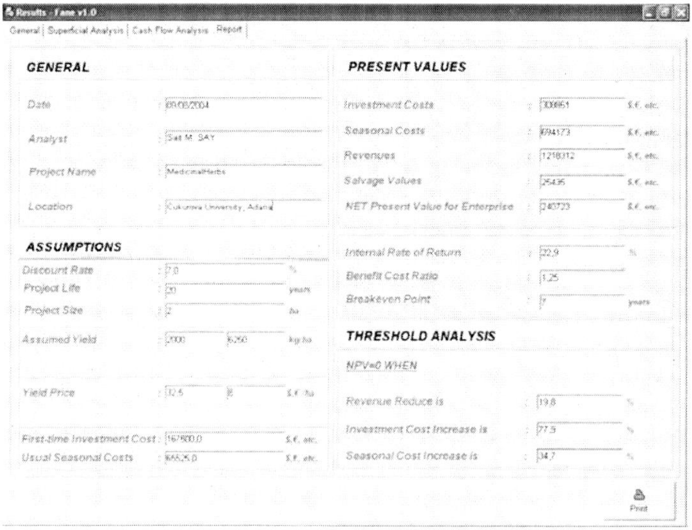

FIGURE A.12. Printable Report screen.

revenue items are valued on a yearly basis over the project life in these subsections. Also, present values for all cost and revenue items are calculated.

A second section of the Cash Flow Analysis Result screen is Result Evaluation. In the Result Evaluation section, Net Present Value, Internal Rate of Return, Benefit–Cost Ratio, and Breakeven Point of the project are presented. Also, Sensitivity (on NPV, IRR) and Threshold Analyses (on revenue decrease, investment cost increase and seasonal cost increase) are performed (Figure A.11).

The last output screen of the software is Report Page. The printable Report Page contains the collection of entered and calculated data, which are needed to decide whether to invest in the proposed enterprise or not (Figure A.12).

Index

Page numbers followed by the letter "f" indicate figures; those followed by the letter "t" indicate tables.

Absolutes, 233, 234f, 235
Acid rain, 18
Active ingredients. *See also* Essential oils; Extraction
  concentration, 4, 8
  location, 10, 95, 253f
  and microwave drying, 95
  sterilization impact, 89, 91
  temperature effects, 97-99
  threshing effect, 143
Adhesives, 261
Adsorbents, 243
Adsorption isotherm, 2
Aflatoxins, 9
Agricultural practices
  improvement, 10-16
  information, 274
  management, 268
  product improvement, 281
  records, 275
  resource allocation, 276, 276f
  uncertainty, 276-279, 277f
Agricultural uses, 260-262
Air velocity, 103, 104
Airflow. *See also* Ventilation
  and classification, 193-198
  and drying, 108, 109f
  and separation, 138f, 139, 141-142, 172, 175, 188, 193-198, 195f, 196f, 197f, 199f, 200f, 201f
Alpine lady's mantle, 54, 55f
Angelica, 241
Animals
  as contaminant, 9, 12, 13
  food for, 17
Anise, 241
Apple rose, 73, 74
Aroma, 244
Aromatherapy, 259
Arsenic, 9
Ash, 54
Asparagus, 35f
Aspergillus, 261
Azinphosmethyl, 18

Bacteria. *See* Microbes
Band dryers, 111-114, 113f, 114f
  automated features, 117
  and separation, 138
  speed, 118
  with sterilizer, 90
  versus batch type, 89
Bark, 17, 224, 228
Basil, 241, 260
Batch dryers, 89, 104-107, 105f, 106f, 107f, 115, 116, 117f
Benefits/cost ratio, 292, 293t, 306f, 307
Bergamot, 230, 241, 261
Berries, 74, 74f
Biogas, 125
Blackthorn, 47
Branches, 60, 73-75, 74f
Breakeven analysis, 295, 295t, 306f, 307
Breeding, 72
Bulbs. *See* Roots
Bundles, 32, 33f
Butterbur, 14

Medicinal and Aromatic Crops

doi:10.1300/5895_09

Cabinet dryers, 106, 107f, 122
Cadmium, 9, 19
Caraway, 76. *See also* Seeds
Carbon dioxide extraction, 236-243, 237f, 238t, 240f
Cardamom, 241
Cascade airflow separators, 196, 199f
Cash flow analysis, 290-292, 290f, 306f, 307f
Chamomile
  contamination, 14
  cutter, 164
  drying, 95, 98, 102
  essential oils, 98, 102
  extraction, 224, 241
  harvesting, 29-30, 31f, 32f, 57f, 58f, 59f, 60-67, 63f, 64f, 65f, 67f, 71f, 72, 72t
  separation, 140-142, 140f
  sterilization, 89
Chemical additives, 85, 133
Chromatography, 239
Cinnamon
  extraction, from leaf, 223, 241
  uses, 260, 261
Citronella, 223
Citrus flavors, 257
Citrus oils, 230, 231, 241, 243, 261
Classification. *See also* Separation
  and airflow, 193-198
  coefficient of friction, 198, 200, 201f
  manual versus mechanized, 170-172
  MAP criteria types, 2t
  by size, 173-193
    and sieves, 173-177
Cleaning
  mixers, 204
  plants, 132-137, 133f, 134f, 135f, 136f, 137f
    fruits, 74, 74f
    and price, 11
    roots, 11-12, 39, 132, 133f, 134f
    seeds, 198-199, 202
  sieves, 191-192, 193f, 194f
  water, 132-134
Cloves, 223, 241, 261
Codex Alimentarus, 7
Coefficient of friction, 198, 200, 201f
Cold fat, 234
Cold pressing, 230
Color
  and arsenic, 9
  and drying, 53, 95, 96, 98, 99, 100, 101t, 118
  and harvesting, 55
  and threshing, 144
Coloring, 257
Coltsfoot, 14
Combined harvesters
  comparisons, 72t
  conveyors, 44
  for flax, 81f
  for flowers, 52, 53f, 55f, 57f, 70f
  for fruits, 73, 74f
  for grains, 76-79, 77t, 81f
  horseradish type, 38, 40f, 41f
  for leaves and stalks, 44, 47-48, 47f, 48, 49f, 139
  for pumpkin seeds, 80, 80f
  self-propelled, 64f, 65f, 69f
  separation, 67-68, 139
  sieve included, 68-70, 69f, 70f, 146
  threshing included, 146
  tractor-mounted, 66-67, 68f, 74f
  tractor-trailed, 74, 74f
  trailed, 50, 50f, 51f, 52f, 67, 69, 70f
Combs, 30, 32f, 52, 53f, 58-69, 58f, 59f, 60f, 61f, 62f, 63f, 64f, 65f, 66f, 67f, 68f
Concretes, 233, 234f
Consumer trend, 6
Contamination. *See also* Microbes
  decontamination, 9, 88-91
  during drying, 105
  during washing, 132-133
  with metals, 9, 18-19
  with plants, 14, 18, 27
  with pollutants, 17-19
  risky practices, 12-13
  safety and quality, 7-10
Convection heat transfer, 93, 95, 97-103
Conveyors
  flowers, 61, 65f
  leaves and stalks, 44, 45f, 48, 50, 50f, 53, 54f, 77, 135
Copper, 9, 19
Coriander, 76, 241
Corrosion, 245
Cosmeceuticals, 259
Cosmetics, 258
Cost annualization, 286

Cost/benefit ratio, 292, 293t, 306f, 307
Costs
  annualization, 286
  cost/revenue, 284-290, 285t, 286t, 287f
  of credit, 278, 286-288
  cultivated versus wild craft, 14
  drying
    dryer type, 105, 107-108, 111
    freeze drying, 97
    fuel, 115, 122, 123, 125-127
    heat exchanger, 113
    microwaves, 96
    temperature, 100
  extraction, 213f, 227, 234, 240, 244
  harvesting, 48, 51, 53, 74
  information, 275
  of mechanization, 19, 29, 48, 51, 53
  seasonal, 285t, 286t, 288, 307
  of sieves, 186, 190
  sterilization, 90
  summary, 285t
  transport, 83
  washing, 134
Credit, 278, 286-288
Critical velocity, 193-195
Crumbling, 156, 156f
Crushers, 164-166, 167f, 168f
Crushing, 156, 156f
Cultivation. *See* Agricultural practices; Wild crafting
Cumin, 241
Cutter bars
  conventional, 50, 50f
  double-knife, 43, 45f, 46f, 48, 51, 52-53
  for grain harvest, 76, 78
  torpedo effect, 45f
Cutters
  exact-cut, 157-160, 158f, 159f, 160f
  feeding mechanism, 159, 160f
  free-cut, 160-164, 161f, 162f, 163f, 164f, 165f, 166f
  and separation, 139
Cutting, 154-156, 155f
Cypermethrin, 18

Dandelion, 241
Decision making
  guidelines, 269-272, 271f, 277f
Decision making *(continued)*
  for investment, 283, 283f, 291
  with software, 303-305, 304f
Demand, 5-6, 16, 254-259, 285t
Density, 96, 201, 214
Desorption isotherm, 86
Deterpenation, 231, 238, 243
Diffusion, 238
Discounting, 291
Distillation
  by steam, 224-230, 225f, 226f, 229f
    hydrodiffusion, 228-230
    turbodistillation, 228
  water method, 214-216, 214f, 215f
    of rose oil, 216-222, 217f, 218f, 220f, 221f, 222f
  water-steam, 223, 223f
Diversity, 3, 5
Dog rose, 73, 79, 168
Drag coefficients, 193, 194
Dragonhead, 241
Drum washers, 135, 135f, 136f
Drum wrapping, 50, 76
Dry-cleaning, 132-137, 133f
Dry processing, 131
Dryers
  batch type, 104-116, 105f
    automated, 116, 117f
    box shape, 105, 106f, 108, 108f, 115, 116, 117f
    cabinet, 106, 107f, 122
    versus band, 89
  continuous (band), 89, 111-114, 113f, 114f, 115, 118
    automated features, 117
    and separation, 138
    sterilizer, 90
  grower sharing, 115
  selection, 114, 120-122
  semi-continuous, 107-111, 109f, 110f, 111f, 121f, 138-139
  solar-based, 119-123, 121, 122, 126
  throughput, 87
Drying
  air velocity, 103, 104
  bundles, 32, 33f
  collection after, 27
  convection heat transfer, 93, 95, 97-103
  costs (*see under* Costs)
  elimination, 244

Drying *(continued)*
- energy for (*see* Energy)
- evaluation, 87
- freeze drying, 96, 98
- and humidity, 86, 93, 101-103, 116
- indirect, 115
- microwave/hot air, 95
- microwaves, 94-96
- monitoring, 115-119
- natural method, 91-92, 91f, 92f
- on-field, 78, 81-82, 119, 120f, 121f
- and pH, 97
- poor practices, 12, 13
- preliminaries (*see* Separation; Washing)
- retention time, 88
- roots, 104, 112
- safety measures, 88-91, 95
- temperature (*see under* Temperature)
- time (duration), 94, 118, 121f, 139, 151-152
- timing (when), 55-56, 89, 94, 97, 115, 123

Dust, 196, 198f, 202-203, 204f
Dyes, 261

Echinacea, 14-15
Ecosystem. *See* Environmentalism
Energy
- for cutting, 154-155
- for drying
  - after surface spraying, 135
  - dryer types, 104, 105, 107, 109, 111-113
  - energy consumption, 87, 101
  - fuel selection, 115
  - liquid biomass, 125, 127
  - monitoring, 118
  - at night, 123
  - predrying separation, 139
  - solar energy, 119-123, 120f, 121f, 122f
  - solid biomass, 123-125, 124f, 126, 127
  - and threshing, 152
- for extraction, 244, 245, 246
  - for distillation, 215, 218, 221f, 224, 227-228

Energy *(continued)*
- for milling, 167
- PTO power, 50, 67, 69
- for sterilization, 90

Enfleurage, 234
Entrepreneurism
- decision-making, 270-272, 271f
- enthusiasm, 273
- information, 274-275, 280, 281
- marketing, 279-281, 280f
- profitability, 282-283 (*see also* Financial analysis)
- resources, 275, 276f, 296
- and risk, 276-279, 277f, 295

Environmentalism
- and extraction, 238, 243
- habitat loss, 5, 11
- overexploitation, 11, 16
- pesticides, 260
- pollution, 13, 17-19, 124
- renewable energy, 119-127
- threatened species, 11, 17

Equilibrium moisture content (EMC), 86, 112, 117
Equilibrium relative humidity (ERH), 86
Ergonomics, 20, 31f, 32-34, 35f, 50, 239
Escherichia coli, 9
Essential oils. *See also* Extraction
- concretes and absolutes, 233, 234f, 235
- definition, 211, 212
- loss
  - chopping, 227
  - distillation, 215, 227
  - drying, 89, 95, 98, 99f, 100t, 118
  - harvesting, 53, 55, 78t
  - milling, 167
  - pre-drying separation, 139
  - size reduction, 227
  - sterilization, 89-90
- plant location, 211-212
- quality, 98
- volatility, 227, 244

Esters, 125, 127, 227
Ethylene oxide, 9
Eucalyptus, 223, 241, 261
European Grower Association, 11, 13-15
European Union, 9
Exact-cut cutters, 157-160, 158f, 159f, 160f

Expenses, 285t
Exports, 16, 24
Expression, 230
Extraction
cold fat, 234
cold pressing, 230
definition, 212
distillation
steam, 224-230
turbodistillation, 228
water, 212-222
water-steam, 223
by maceration, 235
microwave, 245
and particle size, 241
phytol, 246
by solvents, 231-243
sonication, 235
supercritical CO_2, 236-243
and stems, 72-73
subcritical water, 243-245

FANE software, 296
decision screen, 303-305, 304f
flow charts, 300f, 301f
general screen, 300
investment screen, 301
new Project screen, 300, 302f
report creation, 307f
results screen, 305-307, 305f, 306f
revenues screen, 302
start screen, 299
Fennel, 76, 78f
Fermentation, 13
Fertilizer, 12, 18
Financial analysis
breakeven point, 295, 295t, 306f, 307
cash flow, 290-292, 290f, 306f, 307f
cost annualization, 286
cost/revenue, 284-290, 285t, 286t, 287f (*see also* Costs)
goals, 282, 296
internal rate of return, 292
performance indicators, 288t, 291-293, 293t
ratio calculations, 288, 292
Financial analysis *(continued)*
real interest rate, 286-288
sensitivity analysis, 293, 294t, 306f, 307
software for, 296 (*see also* FANE)
superficial stage, 284-288, 287f, 305f
threshold analysis, 289f, 294, 295t, 306f
Financial risk, 278
Fire, 95, 116
Flavorings, 2, 212, 246, 256-257. *See also* Herbs; Spices
Flax oil, 79-80
Floral water, 214
Florasol, 246
Florentine vessel, 214, 219, 222f, 226f
Flotation test, 12
Flowers. *See also* Chamomile
drying, 92f, 104, 107
extraction, 234, 241
harvesting, 34f, 52f, 53f, 56-73, 60f, 61f, 62f, 66f, 68f, 69f, 70f, 216, 217f
stripping, 56
quality, 34, 44
separation, 140-142, 142f
stem cutting, 164, 165f, 166f
Free-cut cutters, 160-164, 161f, 162f, 163f, 164f, 165f, 166f
Freeze drying, 96-97, 98
Frictional drag coefficient, 194
Fruits
citrus, 230, 241 (*see also* Citrus flavors; Citrus oils)
drying, 107, 112, 120, 121f
as fuel, 124, 127
harvesting, 73-75, 74f, 79
holding forces, 73t
pericarp/needles, 168
separation, 168
sliced, 107
Fuel. *See* Energy
Functional foods, 256

Geographic distribution, 4t
Geranium, 241
Germination, 14

Ginger, 97, 241
Gland hairs, 89, 95, 97, 99
Globe artichoke, 90
Governmental role, 15, 21
Grains
  drying, 104, 107, 112, 118, 119
  harvesting, 75-79, 77t, 81f
  pesticides, 260
  separation, 198, 200
  sieves for, 181f, 187
Green forage, 47, 48f, 50, 52f, 154, 157, 158f
Green house, 122, 122f, 126
Grinding
  leaves and stalks, 148, 150t
  spices, 169, 171f
Gross margin, 288, 289
Growers
  association, 16
  mechanization, 19-23. 22f (*see also* Mechanization)

Habitat shrinkage, 5
Harvesting
  and bacteria, 9
  flax oil, 79-80
  flowers, 52f, 53f, 56-73, 60f, 61f, 62f, 66f, 68f, 69f, 70f
  fruits, 73, 79
  of grains, 75-79, 77t, 81f
  by hand, 29-31, 30f, 31f, 33f, 34f, 73-75, 79, 216, 217f
  and humidity, 30, 88, 93
  irrigated plots, 55f, 56, 56f
  leaves and stalk, 40-56
  leaves only, 38f, 54-55
  lodging, 78, 79f
  low plants, 43, 45, 54, 55f
  mowers, 17, 21, 21f
  pumpkin seeds, 79-80
  pumpkins, 52-54, 54f
  roots and bulbs, 37-40, 38f, 39f, 40f, 41f, 42f
  semi-mechanization, 32, 35f, 36f, 79
  success factors, 27-28, 70, 75-77
  temperature, 30, 93
  timing, 12, 28, 30, 54, 77, 78f, 83
  two-phase, 54, 78f
Hazard Analyses and Critical Control Points (HACCP), 7-8
Herbicides, 9, 12, 18
Herbs
  definition, 1
  drying, 107, 111
  extraction, 241
  plant parts, 2
  surface spraying, 135
  in tea, 257
Holding forces, 73t
Hoppers, 44, 45f, 49f
Hops, 95, 108, 109f, 241
Horseradish, harvester, 38, 40f, 41f
Hot air drying, 93
Humidity
  and drying, 86, 93, 101-103, 116
  and harvesting, 30, 88
Hydraulics, 44, 46f, 50, 51
Hydrofluorocarbon, 246
Hydrosol, 214
Hyperforin, 14
Hyssop, 150t

Industry support, 16
Information, 274, 280, 281
Information technology, 6, 296
Infra red, 118
Infusion, 235
Insecticides, 260
Insects, 12, 17, 93
Inspection, 170, 203
Interest costs, 278, 286-288
Internal rate of return, 292, 293t, 306f, 307
Investment, 16, 24. *See also* Costs; Financial analysis
Iris, 241
Irradiation, 9
Irrigation
  and contamination, 12, 13
  and harvesting, 55f, 56, 56f

Jasmine, 233, 235, 241
Javanese pepper, 97
Juniper, 241

Labels, 14
Labor
  distillation, 215
  drying, 104, 109, 111, 112, 116, 127
    surface spraying, 135
  ergonomics, 20, 31f, 32-34, 35f, 50, 239
  extraction, 215, 234, 246
  harvesting, 29-30, 67, 68, 73, 216
  inspection, 203
  social aspects, 34
Laths, 146, 147, 149f
Lavandula, 241
Lavender, 241
Lead, 9, 19
Leaves and stalks
  drying, 96, 107, 112
  harvesting, 40-56
    color change, 53, 55
    leaves only, 38f, 53, 54-55
  leaf grinding, 148, 150t
  separating, 138-140, 138f, 140f, 141f, 143-154, 187-188, 196, 198f
  size reduction, 170f
  stalk percentage, 139
  threshing losses, 148-150, 152
  washing, 135
Lemon balm, 47f, 54-55
Lemongrass, 223
Lifters, 44, 46f, 50
Lipids, 243
Lodging, 78, 79f
Low plants, 43, 45, 54, 55f

Maceration, 235
Management
  decision-making, 269-272, 271f, 277f
  definition, 268
  enthusiasm, 273
  flow chart, 275f
  information, 274, 280, 281
  marketing, 279-281, 280f
  profitability, 281, 282-283 (*see also* Financial analysis)
  record keeping, 275
  resources, 275, 276f, 296
  and risk, 276-279, 277f, 295
Manganese, 9
Manual harvesting, 29-31, 30f, 31f, 33f, 34f, 73-75, 79, 216, 217f
Manual inspection, 170, 203
Manual threshing, 145, 152t
Manufacturers
  quality role, 15-16
  requirements, 19-23
Marigold, 30, 72
Marjoram, 89, 90, 95, 241, 244
Market. *See also* Uses
  for bundles, 32
  competition, 3, 267, 269
  evaluation, 280-281, 289
  risk factors, 278
  supply and demand, 5-6, 16
  and yield, 6
Market channels, 281
Market research, 280
Marketability, 5
Marketing, 16, 279-281, 280f
Marshmallows, 163, 165f
Mechanization
  cleaning, 132-137, 133f, 134f, 135f, 136f, 137f
  cost, 19, 29, 48, 51, 53
  of drying (*see* Dryers)
  of extraction (*see* Extraction)
  flow chart, 22f
  of harvesting
    flowers, 52f, 53f, 56-73, 60f, 61f, 62f, 66f, 68f, 69f, 70f
    fruits, 73
    grains, 75-79, 77t, 81f
    leaves and stalk, 40-56, 139
    roots and bulbs, 37-40, 38f, 39f, 40f, 41f, 42f
  implementation, 19-23
  of inspection and classification, 170
  of sieve cleaning, 191-192, 193f
  of tea packaging, 207f, 208, 208f
  of threshing, 152-154, 152t
Medicinals, 4, 79, 80, 254-255
Mentha, 18, 19. *See also* Peppermint
Metals, 9, 18-19
Methanol, 242
Microbes
  aflatoxins, 9
  and drying, 85, 86, 88, 96, 117
  during transport, 82, 82f

Microbes *(continued)*
  and prices, 11-12
  protectors against, 260
  and quality, 9
  testing, 12
  and washing, 11-12, 134
Microwaves
  drying, 94-96, 118
  extraction, 245
  sterilization, 90
Milk thistle, 76, 77, 120f
Milling, 157, 157f, 167, 170f, 172f
Minerals, 9, 19
Mixing, 204-205, 206f, 227
Moisture
  and ambient air, 86-87, 87f, 101-103, 108
  and band dryers, 112, 113
  and batch dryers, 104
  equilibrium, 117-119
  and essential oils, 98, 99f
  of grains, 75, 76-78
  and microwave drying, 95
  and temperature, 86, 87f, 98
  threshing issues, 150, 151-152
  and transport, 82
  wet-basis, 85-86
Mold
  inhibitors, 261
  and moisture, 86
  and quality, 9
  and relative humidity, 93
  and transport, 82
Mower loaders, 50, 51f
Mowers
  for grains, 54f, 78
  for leaves and stalks, 54
  problems, 43-44
  self-propelled, 47f, 48f, 49f
  for small plots, 54, 55f
Mushrooms, 107
Mustard, 78-79, 79f

Needles, 168
Net margin, 288, 288t, 289
Net present value (NPV), 291, 293, 293t, 295t, 306f, 307
Nets, 91, 92f
Nitrate, 5
Nongovernmental organizations (NGOs), 15, 17, 22f
Nutmeg oil, 260

Onions, 241, 260
Orange blossoms, 233
Oregano, 147f, 241, 260, 261
Organoleptic traits, 144
Origanum dubium, 147f
Overall drag coefficient, 194

Packing, 204-208, 205f, 207f, 208f, 281
Paddles, 62, 77
Paprika, 241
Parsley, 42, 95
Particle movement equation, 175
Patchouli, 223
Pellets, 124
Pepper, 97, 108, 109f, 241
Peppermint
  color, 53, 99, 101t
  drying, 53, 95, 99, 101t
  extraction, 223, 224, 241, 244
  harvesting, 53, 54-55
  separation, 139
  sorption isotherms, 87
  threshing, 147f, 150, 150t, 151
  uses, 261
Perennials, 44
Perfume, 258
Pericarp, 168
Pest control
  pesticides, 9, 18, 260
  as use, 260
Petasis hybridus, 14
pH, 97
Phytol, 246
Phytoremediation, 18, 19
Pickers. *See* Combs
Piper retrofractum. See Javanese pepper
Plant growth regulators, 262
Plant height, 70
Plant oil esters, 125, 127
Plant shape, 96
Plucker-pruner, 28, 29, 30
Polarity, 242, 243

Pollution, 13, 17-19, 105, 124
Pomade, 235
Poppy seeds, 77
Potato–onion digging machines, 39, 42f, 43f
Power-takeoff (PTO) power, 50, 67, 69
Pressure, 227, 235, 236-243, 237f, 238t
Prices
  and cleaning, 11-12
  market factors, 3, 6, 16
  and market research, 281
  and processing, 213f
  as risk factor, 278
Primrose, 241
Processing
  quality, 7-16
  seasonal expenses, 285t
  steps, 3f
  and value, 213f
Profit, 281, 282-283
Prototyping, 19-23
Prunus Africana, 17
Prunus spinosa. See Blackthorn
Pumpkin seeds
  drying, 107, 108, 118
  harvesting, 79-81, 80f
  washing, 135, 136f
Purity, 5, 8-10
Pygeum, 17
Pyrethrum, 30, 68f, 72-73, 260

Quality
  of active ingredients, 10
  and drying
    degradation, 85, 96
    dryer type, 111
    freeze drying, 97
    humidity, 88, 103
    safety, 88-91, 105, 119
    timing, 81-82, 115
  environmental issues, 16-19
  extraction, 215, 224, 227, 239, 244
  and harvesting, 28, 50, 53, 72t, 93
  improvements, 7-16
  inspection, 203
  separation, 141
  and size, 160
  transport, 81-83
  wild crafting, 4

Radioactivity
  as contaminant, 10, 12
  as decontaminant, 9
Rain, 54
Rape, 76
Record keeping, 275
Reels, 43, 45f, 46f, 47, 51, 52, 77
Refrigerant, 246
Relative humidity
  and drying, 86, 93, 101-103, 116
  and harvesting, 30, 88
Resinoids, 232
Resource management, 275, 276f, 296. *See also* Energy
Retention time, 88
Return on Investment (ROI), 288
Return on seasonal costs, 288
Revenues, 286t, 302, 304f, 307
Ring-barking, 17
Risk
  in business, 276-279, 277f, 295
  of contamination, 12
Rolling resistance, 198, 201, 201f, 202f
Roots
  cleaning, 11-12, 39, 132, 133f, 134f
  drying, 104, 112
  extraction, 224, 228, 241
  harvesting, 37-40, 38f, 39f, 40f, 41f, 42f
  shape variations, 39f
Rosa canina. See Dog rose
Rosa villosa. See Apple rose
Rose flowers, 241
Rose oil
  extraction, 233, 235
    distillation, 216-222, 217f, 218f, 220f, 221f, 222f, 224
Rose water, 220
Rosemary, 95, 241

Safety, 7-10, 12-14, 17-19
Saffron, 34f, 141, 142f, 191, 192f
Sage
  drying, 98-99, 99f, 100t, 101, 102
  extraction, 241
Sand, 186, 195
Sandalwood, 241
Savory, 241
Sea buckthorn, 73

Seasonal inputs, 285t, 286t, 288
FANE software screen, 301, 302f, 303f, 307
Seeds. *See also* Pumpkin seeds
extraction, 224, 228, 241
harvesting, 76-77
separation, 198, 200, 202
threshing, 152
Semi-continuous dryers, 107-111, 109f, 110f, 111f
Sensitivity analysis, 293, 294t, 306f, 307
Separation
after drying, 143-154, 168
by airflow, 138f, 139, 141, 172, 175, 188, 193-198, 195f, 196f, 197f, 199f, 200f, 201f
before drying, 67-69, 69f, 137-142, 138f, 140f
coefficient of friction, 198, 200, 201f
distinguishing traits, 173, 173f
of dust, 196, 198f, 202-203, 204f
electrical properties, 202, 204f
by gravity, 199-202, 203f
on harvester, 67-68, 69f
of pumpkin seeds, 80, 80f
by rolling resistance, 198, 201, 201f, 202f
of soil, from roots, 39
by stem length, 67
Shaking, 74, 75f, 181-188, 181f, 186f, 187f
Shape, 39f, 96. *See also* Low plants
Shelf dryers, 109-111, 111f, 122, 138-139
Sickle blades, 54
Sieves
cleaning, 191-192, 193f, 194f
for cleaning, 132
cost, 186, 190
in crusher, 166
in cutters, 161, 162f, 163f, 168
cylindrical (drum), 140-141, 140f, 141f, 188-190, 191f, 192f
groepel type, 146, 146f
in harvester, 68-70, 69f, 70f, 146
and milling, 157, 157f, 167, 168
oscillation, 181-188, 181f, 186f, 187f
particle movement, 174-177, 176f
Sieves *(continued)*
round, 186, 188f, 189f
separation, 177, 178
and size classification, 173-178, 174f, 176f
square, 187
and threshing, 143, 145, 146f, 148, 149f
vertical and horizontal, 182-185, 183f, 184f, 185f, 189f, 190f
Silica gel, 243
Site selection, 12
Size
classification, 173-193
particle distribution, 178, 179f
particle size, 178
and extraction, 241
and mixers, 205
of plot (*see* Small plots)
reduction
crushing, 156, 156f, 164-166, 167f, 168f
cutting, 154-156, 155f, 157-164
and essential oils, 227
fruits, 168
milling, 157, 157f, 167, 170f, 172f
of spices, 169, 171f
and threshing, 143, 144, 147, 153f
Small plots
with irrigation, 55f, 56, 56f
mowers, 54, 55f
Software, 296
Soil
compression, 44
and leaf/stalk harvesting, 54
removing, 11-12, 39, 132-137, 133f, 134f, 135f, 136f, 137f
residues in, 18
Solar energy, 119-123, 120f, 121f, 122f, 126
Solar radiation, 93
Solvents, 231-233, 234f, 237-243
co-solvents, 242
and microwave method, 245
organic, 232-234
sonication, 235
Sorption isotherm, 86, 87f
Spearmint, 224, 241
Specific energy, 87, 101, 103, 104

Spices
definition, 2
extraction, 241
grinding, 169, 171f
St. John's wort, 14, 88, 95
St. Mary's thistle, 241
Stability, 144, 231. *See also* Humidity; Moisture; Terpenes
Standardization, 7, 11, 28, 85-86
Steam
extraction, 224-230, 225f, 226f, 229f
sterilization, 89
Stems
cutting, 164, 165f, 166f
harvesting, 30, 60, 66, 67-70, 72-73, 73t
of chamomile, 44
Sterilization, 9, 89-91
Stones, 195
Storage
of citrus oils, 231
of drying trays, 93
of plant material, 13, 85, 204-208
Straw, 75, 76, 127, 198
Stripping, 34, 36f, 56, 62-64, 66f
Subcritical water, 243-245
Supercritical CO_2, 236-243, 237f, 238t, 240f
Superheated water, 243-245
Surface tension, 238
Swathes, 43, 54f
SWOT analysis, 280

Tea. *See also* Chamomile
cutting, 166
decaffeination, 240
harvesting, 53, 54, 55f
ingredients, 257
packaging, 205-208, 207f
Temperature
for drying
color impact, 95, 99, 99f, 101t
essential oils loss, 100t
freeze drying, 96
hot air drying, 93, 95, 97-99
quality impact, 87-89, 115
and specific energy, 101
straw energy, 123
Temperature *(continued)*
and extraction, 236-246, 237f, 238t (*see also* Distillation)
and harvesting, 30, 93
and transport, 82, 82f
Terminal velocity, 193
Terminology, 1
Terpenes, 231, 238, 243
Threshers, 74f, 145-147, 145f, 146f, 147f, 148f, 149f, 150f, 151, 153f
Threshing
of caraway, 76
of coriander, 76
desirable traits, 144
friction issues, 143-145
manual, 145, 152t
mechanized, 152-154, 152t
post-drying, 143-154
pre-dried material, 53
Threshold analysis, 289f, 294, 295t, 306f
Thyme, 42, 43, 96, 150f, 150t, 223, 241, 260, 261
Timing
drying
duration, 94, 118, 121f, 139, 151-152
in processing sequence, 70, 81-82, 89, 94, 97, 115, 123
extraction
distillation, 219, 224, 227, 228
new methods, 239, 242, 243
harvest, 12, 28, 30, 54, 77, 78f, 83
inspection, 203
leaf removal, 151
separation, 67-69, 69f, 137-143, 138f, 151
sterilization, 90
transport, 82-83
Tobacco, 158
Tools. *See* Combs; Cutters; Mechanization; Stripping
Torpedoes, 44, 45f, 46f, 50
Toxicity, 12-14, 17-19
Tractors, 50-51, 66-67, 68f, 74, 74f, 80
Trailed machines, 50-51, 50f, 51f, 52f, 67, 69, 70f, 74, 74f
Transport. *See also* Conveyors
contamination, 13, 18, 82, 82f, 88
duration, 82

Transport *(continued)*
during harvest, 34, 44, 50-51, 61, 67, 88
pumpkins, 80
market research, 281
packing, 204-208
and threshing, 147
time of day, 82-83, 82f
of washed herbs, 135
Trays, drying on, 03&8f, 91-93
Trichomes, 89, 95, 97m 15
Tuberose, 235
Tunnel dryers, 109-111, 111f, 120, 121f

Ultrasound, 235
Uses, 2-4, 256-262
by country, 4t
as medicine, 4, 254-255

Valerian, 40, 43f, 131-132, 241
Value-added, 281
Ventilation, 34, 36f, 62, 83, 91, 103
Viscosity, 238

Washers
brush-type, 136, 137f
drum type, 135, 135f, 136f
gravity flotation type, 136, 137f
nozzles, 136, 136f
Washing, 11-12, 132-137, 133f, 134f, 135f, 136f, 137f
Water
for extraction, 228, 243-245
for irrigation, 13
on washed herbs, 135
for washing, 11-12, 132-134
Water extraction, 243-245. *See also* Distillation
Waxes, 243
Weather
and drying, 88, 115, 120-122
and harvesting, 54
Wild crafting
advantages, 14
equipment, 34
overexploitation, 16
regulation, 15
statistics, 4
Windrower, 53, 54f, 78
Wood, 124, 127. *See also* Bark
World Health Organization, 7, 255

Yarrow, 72, 241
Yeast, 86
Yew, 241
Yield, 285t, 286t
Young plants, 44

Zigzag airflow separators, 197, 200f, 201f
Zinc, 9, 19